SpringerBriefs in Computer Science

More information about this series at http://www.springer.com/series/10028

Patrick Laube

Computational Movement Analysis

 Springer

Patrick Laube
Institute of Natural Resource Sciences
Zurich University of Applied Sciences
Wädenswil
Switzerland

ISSN 2191-5768   ISSN 2191-5776   (electronic)
ISBN 978-3-319-10267-2   ISBN 978-3-319-10268-9   (eBook)
DOI 10.1007/978-3-319-10268-9

Library of Congress Control Number: 2014946736

Springer Cham Heidelberg New York Dordrecht London

Printed on acid-free paper

Springer is part of Springer Science+Business Media (www.springer.com)

For Esther and André Laube

Preface

This volume was initially written to fulfill the formal requirements for achieving *venia legendi* (Habilitation) from the Faculty of Science at the University of Zurich, Switzerland. Therefore, the text reports on my research from the years 2006 to 2014. It does so by offering a synthesis from a set of publications I have contributed to on the topic of movement analysis in the wider area of Geographic Information Science. These publications include articles published in peer-reviewed journals and peer-reviewed conference proceedings, as well as conference contributions, book chapters, editorials, and review articles.

The book proposes the theoretical underpinnings of the emerging interdisciplinary research field termed here *Computational Movement Analysis* (CMA). The book features three core chapters, each set out around a number of overarching research questions covering a certain aspect of CMA. The contribution of each chapter is twofold.

First, each of these chapters offers a comprehensive synthesis of work that I was involved in. Emphasizing that the chapters first and foremost summarize my own work, the respective publications are marked with a bold label (**P2**. Laube and Dennis 2006) added to their citation (see Fig. 1.1 for the respective codes). Even though the arguments in the synthesis sections are supported mainly from publications I contributed to, some complementary references helped to complete the picture in these synthesis sections.

Secondly, in order to offer the reader a wider perspective than could be possibly offered through a summary of my own work alone, every chapter concludes with a comprehensive review of the selected related work in the area. These "Related Work" sections systematically revisit the lines of argument in the before made synthesis and discuss further work or alternative approaches in the respective areas.

Furthermore, I happily acknowledge that my personal contribution to the set of publications building the core of this book varies. In many projects I acted as the principal investigator, in others, however, I contributed less. Hence, clearly, the minds of many colleagues helped in shaping the ideas summarized in this SpringerBriefs volume. However, the presented book here offers my synthesis of the developing field of Computational Movement Analysis.

Finally, the choice of the term "Computational Movement Analysis" was inspired by my attending of the Dagstuhl Seminar 10121 on Computational Transportation Science at Schloss Dagstuhl, 21–26 March 2010, and additional fruitful discussions with Joachim Gudmundsson and Thomas Wolle when writing a section for the 2012 Springer Handbook of Geographic Information (**P17**. Gudmundsson et al. 2012).

Zurich, June 2014   Patrick Laube

References

Gudmundsson, J., Laube, P., & Wolle, T. (2012). Computational movement analysis. In W. Kresse & D. M. Danko (Eds.), *Springer Handbook of Geographic Information* (pp. 423–438). Berlin Heidelberg: Springer.

Laube, P., & Dennis, T. (2006). Exploratory analysis of movement trajectories. In GeoCart,. (2006).*National Cartographic Conference*. Auckland, NZ.

Acknowledgments

Many people have made this work possible. I would first and foremost like to thank my colleague, mentor, and friend Robert Weibel for the support, freedom, and feedback he has given throughout the completion of this project. Then I thank all colleagues and friends at the Department of Geography for providing a congenial work environment. Special thanks go to Ross Purves for his continuous support and for stimulating and fruitful collaborations on many aspects relating to *Computational Movement Analysis*. I also thank Felix Morsdorf for countless discussion on all matters academia and for a simple yet crucial idea on how to structure the book.

I am indebted to Matt Duckham for introducing me to geosensor networks and for his invaluable support during and after the Melbourne years. I would especially like to thank Matt for acting as a generous host during four very productive weeks during the southern summer in 2013. It was during and after a research exchange visit within the Australian Research Council Discovery Project DP120100072 "From environmental monitoring to management: Extracting knowledge about environmental events from sensor data", when I found at the Department for Infrastructure engineering a hospitable environment for writing large sections of the present book. Also in Melbourne, I thank Marimuthu Palaniswami from electrical engineering, who generously supported my work, and Stephan Winter for useful discussions. Finally, cordial thanks go to the CRC crew, including Franz Rottensteiner, Jochen Willneff, and Jane Inall, for their support and ongoing friendship.

I also thank Pip Forer and Todd Dennis for an inspiring research period in Auckland, studying the movement of New Zealand pigeons, possums, and tourists. Finally, Mark Gahegan and Alan MacEachren from the Geovista Center at Penn State provided hospitality and engaged in useful discussions on data mining and exploratory data analysis.

More help and support came from many quarters. Above all, this includes all my co-authors, especially Joachim Gudmundsson, Thomas Wolle, Marc van Kreveld, Kai-Florian Richter, and Falko Schmid. Equally important contributions to this work emerged from my work with Ph.D. and M.Sc. students, where I'd like to single out joint research with Somayeh Dodge, Christian Gschwend, and Michael Merki.

Finally, I thank Susan Lagerstrom-Fife and Jennifer Malat from Springer for their support and patience when producing this SpringerBriefs volume.

Thank you all.

Contents

Acronyms

CMA	Computational Movement Analysis
DDIG	Deferred Decentralized Information Grazing
DeSC	Decentralized Spatial Computing
EOC	Error of commission
EOO	Error of omission
FLAGS	Flocking Amongst Geosensors
GIS	Geographic Information System
GISc	Geographic Information Science
GPS	Global Positioning Service
GSM	Global System for Mobile Communications
GSN	Geosensor Network
ICT	Information and Communications Technology
KDD	Knowledge Discovery in Databases
LBS	Location-Based Service
MOD	Moving Object Databases
MPO	Moving Point Object
NWED	Normalized Weighted Edit Distance
POI	Point of Interest
QTC	Qualitative Trajectory Calculus
RFID	Radio-frequency identification
VANET	Vehicular ad hoc network
WSN	Wireless Sensor Network

Chapter 1
Introduction

In the last decade, advances in tracking technologies resulted in geographic information representing the movement of individual objects at previously unseen spatial and temporal granularities.[1] This novel, inherently spatio-temporal kind of geographic information offers new insights into dynamic geographic processes but also challenges the traditional very static spatial analysis toolbox (**P17**. Gudmundsson et al. 2012).

Consequently, movement analysis has emerged as a major new research focus of Geographical Information Science (GIScience). I argue that movement is the first truly spatio-temporal phenomenon on geographic scales that is traceable beyond the snapshot. Since movement data is furthermore easily accessible and seemingly simple in structure, its analysis has received increasing attention from the GIScience and wider community.

Work has appeared addressing modeling, storing, indexing, and querying movement, mapping and visualizing movement, movement patterns, trajectory similarity and clustering, trajectory segmentation, semantically annotating and enriching trajectories, as well as simulating movement in the context of many mobile applications, for instance for location-based services (LBS), vehicular ad hoc networks (VANETs), or geosensor networks).

This book has a thesis, it makes the case for Computational Movement Analysis (CMA), as an interdisciplinary umbrella for contributions from a wide range of fields aiming for a better understanding of movement processes. This first chapter explains why this inclusive umbrella is a contribution, what it involves, and which fields it borrows methods and concepts from.

---

[1] There is a large body of literature on movement analysis with a medical kinetics perspective, studying the movement of body parts. Such movement is not covered in this book.

© The Author(s) 2014
P. Laube, *Computational Movement Analysis*,
SpringerBriefs in Computer Science, DOI 10.1007/978-3-319-10268-9_1

1.1 Motivation

Similar to many other research fields, the spatial sciences have made a transition in the last decades from computation and data poor to computation and data rich environments (Miller and Han 2009). The revolutionary character of this transition is especially evident in the application field animal navigation research, where the movement behavior of individual animals has been studied for decades as a fundamental knowledge base. Not so long ago, tracking of moving entities was a very cumbersome and costly undertaking: For example, turtles were tracked using *thread-trailing* (Claussen et al. 1997) and all that flying birds would reveal was their *vanishing bearing* in release experiments (Walker 1998). By contrast, to date it is now common to track flying birds using GPS devices with a sampling rate of seconds (Shamoun-Baranes et al. 2012). Hence, the previous lack of fine-grained movement data is a first reason why CMA is a relatively young and little-developed research field.

Secondly, within GIScience, the legacy of cartography's static view of the world slowed down the development of CMA. For decades, dynamic processes and change was primarily captured in a snap-shot manner, where the arrival of a new areal or satellite image only allowed a comparison with previous states of the world (Worboys and Duckham 2004). Finally, geography's snap-shot view unfortunately found a match in the concept of sporadic updates in the databases underlying early geographic information systems. Conventional databases are designed for handling mostly static data with occasional updates (e.g., a change of ownership or an area change in a cadastral map), but not entities that change continuously (e.g., the permanently changing location of a moving car).

The increased availability of movement data goes hand in hand with a growing interest of the application fields in exploiting that new resource. Be it behavioral ecology, transportation and mobility research, surveillance and security, or even sports analysis, all fields interested in movement show significant interest in studying and analyzing movement. For example, only for the field of movement ecology, Holyoak et al. (2008) list thousands of papers addressing organismal movement, ranging from seed dispersal to bird migration. The question arises if the described data and information revolution requires new scientific foundations with respect to methods. With richer and more complex data comes a need for more sophisticated tools for managing, exploring and analyzing that data. Arguing for the need of new scientific fundamentals CMA first of all means identifying the characteristics of the new challenge. Similar to Anselin (1990) seminal question about "What is special about spatial?", it is now fair to ask "what is special about movement?".

The following list summarizes a set of properties of movement (data) that underline the need for a new theory of computational movement analysis. The list is inspired by revisiting seminal texts framing the theory of GIScience (Anselin 1990; Goodchild 1992, 2001), and adapting the challenges therein to the movement domain.

- *Geographic reference systems.* Movement traces are located in space-time, hence their management and analysis requires a measurement framework of up to three interrelated spatial dimensions.
- *Temporal reference systems and sequence.* On top of the above spatial reference systems, movement inherently requires a fourth, temporal dimension. This dimension is typically directed and keeps progressing, but can alternatively be conceptualized as being cyclical or branching (Frank 1998, 2001). Consequently, observations of moving objects have a sequence that can be time-stamped and typically can't be reversed.
- *Permanent instead of sporadic change.* Moving objects by definition are in motion, and the change of position is the norm, not the exception. This is fundamentally different to conventional spatial objects, where in essence a static world is assumed that undergoes sporadic change or up-date events. Hence, many techniques that rely on supporting data structures (such as indices, trees, aggregations) are fundamentally challenged because the entities permanently rearrange.
- *Movement traces are complex objects.* Whereas the moving objects themselves are most often modeled as simple moving point objects, the traces they leave in space-time are complex. Relations such as distance or similarity are consequently more complex than relations between simple data points in conventional feature space.
- *Implicit relationships.* Just as for relationships amongst spatial objects, relevant relationships between moving objects are often implicit and must be materialized first using metrics and operations considering up to four dimensions. Examples include spatio-temporal topological relations such as meet, line up, diverge or converge.
- *Overlap.* Since gregarious animals and social human beings tend to use similar spaces at similar times, movement traces often cluster in space-time, resulting in overlapping data, clutter, and dense areas with severe information overload.
- *Spatial dependency and heterogeneity.* Positional fixes along a trajectory feature highly autocorrelated attributes. This poses challenges around descriptive statistics, sampling granularities and data compression. At the same time, since movement always happens in a potentially heterogeneous space, movement trajectories are also expected to adopt certain aspects of spatial heterogeneity.
- *Knowledge of movement is inherently uncertain.* Even at finest tracking granularities we can't monitor the complete movement trace of a moving object, hence our discrete view of it must always be uncertain. Uncertainty arises from positional uncertainty of the localization technologies used and the ignorance about what happened in between observed fixes.
- *Most information describing movement is derivative.* Just as geographic information is mostly derivative (Goodchild 2001), so is movement information. Although more and more sensors claim direct sensing of movement properties such as speed or orientation, many measurements describing movement are the result of compilation, calculation, and interpretation, mostly hidden from the user. Even if it was transparent, there are typically many ways of deriving movement descriptors and the reasons for choosing one over the other are often not transparent.

- *Descriptive parameters are inherently scale-specific.* Goodchild (2001, p. 11) makes a very illustrative case that in essence there is "no such thing as the slope of a geographic surface, only a slope *at a specific scale* or grid spacing". Similarly, movement properties such as speed, acceleration or path sinuosity require a sampling scale, and the choice of this scale is in most cases far from being self-evident.

For all those reasons this book argues for the need of new scientific fundamentals for CMA. Following Antony Galton, this new theory shall enable GIScience to bridge the "perceived gulf between, on the one hand, low-level observational [movement] data that constitutes the "raw material" of our science, and on the other hand, the high-level conceptual schemes through which we as humans interpret, understand, and use that data" (Galton 2005, p. 300).

1.2 Introducing Computational Movement Analysis

Computational Movement Analysis (CMA) draws concepts and methods from three methodological research areas: (1) geographic information science or GIScience, (2) computer science, and (3) statistics. Additional important contributions emerge application fields studying movement, such as, for example, movement ecology, surveillance and crowd management, as well as transportation research.

From geographic information science CMA inherits concepts for modeling space and the movement within, as well as a suite of spatio-temporal operations interrelating space, time, and movement. From computer science CMA draws on the database theory on how to store, index, and query inherently dynamic movement data. Also from computer science CMA profits from developments around analytical tools such as data mining, knowledge discovery, and simulation for numerical experiments. Finally, from statistics CMA inherits many techniques for descriptive statistics, exploratory data analysis, and stochastic models for movement simulation (for instance, random walk, states space models). Clearly, these fields overlap. For instance, data mining and visualization reappear in visual analytics approaches for movement data, and mapping spatio-temporal movement requires innovative visualization approaches.

Although most applied research fields studying movement do not explicitly focus on the development of computational movement analysis methods, they still significantly contribute to the respective theory. In movement ecology, for instance, there is a very active community developing statistics based tools for movement analysis. Similarly, many relevant developments for moving object databases emerge an active community addressing fleet management problems.

Definition (*Computational Movement Analysis*) (CMA) is the interdisciplinary research field studying the development and application of computational techniques for capturing, processing, managing, structuring, and ultimately analyzing data describing movement phenomena, both in geographic and abstract spaces, aiming for a better understanding of the processes governing that movement.

CMA investigates the scientific fundamentals related to

- the specific *characteristics* and peculiarities of the geographic phenomenon movement and the spatio-temporal data describing it, including data quality (uncertainty, accuracy), scale issues, and spatio-temporal autocorrelation,
- the peculiarities of established and emerging integrated spatial systems serving as direct or indirect *tracking systems* capturing raw or enriched movement data,
- capturing, (pre-)processing, integrating, storing, managing, and querying the rapidly growing *data* streams describing movement phenomena,
- the *conceptual models* for moving objects and movement processes, and the spaces embedding that movement, the *data structures* implementing these models, and the implications of models and structures on the CMA process,
- the *development and evaluation of analysis techniques and operations* structuring low-level movement data and deriving high-level process knowledge from that data. This draws on methods from spatio-temporal analysis, geography, computational geometry, scientific visualization, data mining and KDD, and statistics.
- the characteristics and *semantics* of the wide range of current *applications* of CMA, and the assessment of the potential of prospective applications areas, and
- *societal issues*, including ethics and privacy, as well as issues around user-generated and open data.

1.3 Structure of this Book

The book is organized into three core chapters, each discussing another aspect of CMA. Chapter 2 investigates the conceptual modeling of movement spaces and the movement embedded in those spaces. Chapter 3 focuses on adopting and adapting data mining techniques for CMA. From the many possible application areas of CMA, Chap. 4 specifically focuses on CMA challenges and opportunities in decentralized spatial information systems. A chapter on grand challenges in the area concludes the book.

Figure 1.1 offers an overview of the publications building core of the book and the topics they cover. In order to give the reader a more detailed overview, I have narrowed down and particularized the generic list of issues given in the CMA definition above, producing a set of specific keywords heading the columns in the pictorial matrix of content in Fig. 1.1. For every included publication I then indicate whether the listed issues build a *key*, *major*, or *minor* topic.

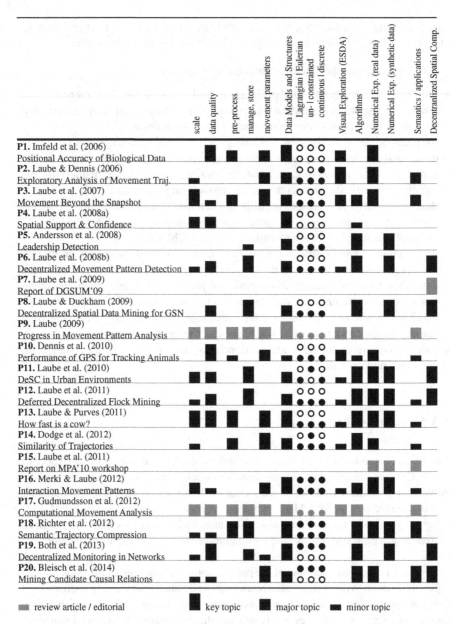

Fig. 1.1 Research articles included in this book and their relative contributions to the scientific fundamentals of Computational Movement Analysis listed in Sect. 1.2

References

Anselin, L. (1990). What is special about spatial data? In D. A. Griffith (Ed.), *Statistics, past, present, and future, monograph Series* (pp. 63–77). Ann Arbor, MI: Institute of Mathematical Geography.

Claussen, D. L., Finkler, M. S., & Smith, M. M. (1997). Thread trailing of turtles: Methods for evaluating spatial movements and pathway structure. *Canadian Journal of Zoology-Revue Canadienne De Zoologie, 75*(12), 2120–2128.

Frank, A. U. (1998). Different types of times in GIS. In M. J. Egenhofer & R. G. Colledge (Eds.), *Spatial and temporal reasoning in geographic information systems* (pp. 40–62). Oxford, UK: Oxford University Press.

Frank, A. U. (2001). Socio-economic units: Their life and motion. In A. U. Frank, J. Raper, & J. P. Cheylan (Eds.), *Life and motion of socio-economic units* (Vol. 8, pp. 21–34)., GISDATA London, UK: Taylor & Francis.

Galton, A. (2005). Dynamic collectives and their collective dynamics. In A. Cohn & D. M. Mark (Eds.), *Spatial information theory, proceedings* (Vol. 3693, pp. 300–315)., Lecture Notes in Computer Science Berlin: Springer.

Goodchild, M. F. (1992). Geographical information science. *International Journal of Geographical Information Systems, 6*(1), 31–45.

Goodchild, M. F. (2001). A geographer looks at spatial information theory. *Spatial information theory* (Vol. 2205, pp. 1–13)., Lecture Notes in Computer Science Berlin: Springer.

Gudmundsson, J., Laube, P., & Wolle, T. (2012). Computational movement analysis. In W. Kresse & D. M. Danko (Eds.), *Springer handbook of geographic information* (pp. 423–438). Berlin: Springer.

Holyoak, M., Casagrandi, R., Nathan, R., Revilla, E., & Spiegel, O. (2008). Trends and missing parts in the study of movement ecology. *Proceedings of the National Academy of Sciences, 105*(49), 19060–19065.

Miller, H., & Han, J. (Eds.). (2009). *Geographic data mining and knowledge discovery* (2nd ed.). Boca Raton, FL: CRC Press.

Shamoun-Baranes, J., van Loon, E. E., Purves, R. S., Speckmann, B., Weiskopf, D., & Camphuysen, C. J. (2012). Analysis and visualization of animal movement. *Biology Letters, 8*(1), 6–9.

Walker, M. M. (1998). On a wing and a vector: A model for magnetic navigation by homing pigeons. *Journal of Theoretical Biology, 192*(3), 341–349.

Worboys, M., & Duckham, M. (2004). *GIS—A computing perspective* (2nd ed.). New York: CRC Press.

Chapter 2
Movement Spaces and Movement Traces

The analysis of the observed movement by means of computers requires abstraction, conceptual modeling, and formalization of the moving entities and the spaces embedding that movement (Peuquet 2002). This preliminary but crucial stage of Computational Movement Analysis (CMA) requires modeling choices but is also constrained by the data sources at hand. This chapter investigates how movement can be modeled from the various data sources contributing to CMA, and discusses implications of the characteristics of models and sources on how movement can be captured and characterized, structured and analyzed.

Overarching research objectives. The research summarized in this chapter contributes to the following overarching research objectives of computational movement analysis.

- Contribute to the establishment of a theory of computational movement analysis, drawing on concepts and methods of GIScience and related research fields.
- Investigate the implications of the conceptual modeling of movement spaces and the movement embedded in these spaces on the process and the outcomes of computational movement analysis.

2.1 Data

This book focuses on the movement of real world entities that can be abstracted as moving point objects (MPOs). The research covered in this book investigates movement of a diverse set of MPOs, including various animal species (for instance, racing pigeons, fish, sheep, cows, and brushtail possums), several expressions of human mobility (including bicyclists, couriers, playing children), movement of abstract objects in the physical environment (hurricanes), as well as simulated movement of software agents (simulated pedestrians, sensor nodes of a wireless sensor network, and agents performing various forms of random walk). Table 2.1 gives a comparative overview of data sources contributing to this volume.

© The Author(s) 2014
P. Laube, *Computational Movement Analysis*,
SpringerBriefs in Computer Science, DOI 10.1007/978-3-319-10268-9_2

Table 2.1 List of diverse MPOs investigated in this book

MPOs	Space	Tracking	δt	Captured for	References
Racing pigeons	EH	GPS	1 s	Avian navigation	Laube et al. (**P3**. 2007)
Sheep	EC	GPS	60 s	Management of domestic animals	Laube and Dennis (**P2**. 2006)
Cows	EH	GPS	0.3 s	Precision farming with WSN	Laube and Purves (**P13**. 2011), Laube et al. (**P12**. 2011)
Brushtail possums	EH	GPS	15 min	Animal ecology	Dennis et al. (**P10**. 2010)
Fish	NET	RFID	24 h	Ecological monitoring	Bleisch et al. (**P20**. 2014)
Bicyclists	NET	GPS	10 s	Wayfinding	Richter et al. (**P18**. 2012)
Couriers in London	NET, AQU	GPS	10 s	Fleet management	Dodge et al. (**P14**. 2012)
Hurricanes	EH	Radar	6 h	Climatology	Dodge et al. (**P14**. 2012)
Simulated pedestrians	NET	ABM	–	CMA	Laube et al. (**P11**. 2010)
Simulated sensor nodes	EH	ABM, random walk	–	CMA	Laube et al. (**P12**. 2011), Laube et al. (**P6**. 2008)
Simulated fish agents	NET	ABM	–	CMA	Both et al. (**P19**. 2013)

Space EH: Euclidean homogeneous, EC: Euclidean constrained, AQU: space-time aquarium, NET: network space. *Tracking* ABM: agent-based modeling, using repast or netlogo. *Captured for* respective application fields or CMA in the first place

The moving objects in Table 2.1 were tracked for a wide range of purposes and using many different tracking technologies. The primary focus of the research covered here is the development of generic methods for the analysis of movement data, and not the understanding of a specific movement process *per se*, with respect to a specific application field. Hence, in no single case was real world movement data captured explicitly, but rather collaborations were sought with researchers collecting data for their own research in movement ecology, environmental monitoring, or transportation research. This is important to note as all real data covered in this book were collected for the specific purposes of the application scientists, and not for the methodological research building the focus of this book.

The evaluation of proposed methods, however, often involved the simulation of movement data under experimental conditions. To that end, synthetic data was repeatedly produced making use of random walk algorithms and agent-based modeling. The advantage of simulated synthetic movement data is that in contrast to real observed movement data, the movement processes to be studied and the hence emerging movement data can be rigorously controlled, which builds a crucial precondition for experimental evaluation of methods (see also Sect. 3.3).

Table 2.1 finally illustrates that even though GPS-tracking still is the most frequent way of tracking MPOs, alternative ways of tracking moving objects have emerged through the technological advancement of location-aware mobile ICT devices.

2.2 Conceptual Models for Movement and Movement Spaces

The above overview illustrates that the modeling of movement means modeling the moving entities, but equally important modeling the space they move in. The varying characteristics of possible conceptual space models embedding a form of movement—be it an animal habitat, a 3D building, or a complex urban transportation network—rules how the entities can move, and consequently impacts on the computational analysis tools that are required and suitable for understanding that movement. Hence, a critical modeling decision early in the CMA process is the choice of the conceptual data model for the space embedding the movement under study.[1]

The review article (P9. Laube 2009) has proposed a categorization of six basic conceptual movement spaces that are commonly found in CMA (see Fig. 2.1). Animals tagged with GPS receivers (e.g., migrating birds) move in an unconstrained Euclidean space (a). Sometimes movement is constrained as non-swimming animals will not enter a lake or in an indoor environment shoppers can't enter locked rooms (b). Especially visualization applications favor the space-time cube metaphor following Hägerstrand's Time Geography (Hägerstrand 1970) (c). Movement can also be captured in discrete tessellations of space, for example, as a series of discrete steps through a field representation of space (d). Location-aware mobile devices leave digital traces as a sequence of visited GSM cells (e). Finally, human movement is often tied to a transportation network, where movement can only occur along edges between intersecting nodes (f).

This categorization proved to be useful for leading the crucial discussion about conceptual data models in CMA, a traditionally data-driven research field often accepting the data-inherent structures as unchangeable preliminaries and neglecting the implications of conceptual design choices. As will be shown in the next chapter about movement mining, the different conceptual movement spaces allow for the detection of different movement patterns (Fig. 3.3). The categorization

---

[1] Note that this section is focused on how movement traces can be abstracted and represented in spatial information systems. Other authors have put forward conceptual models for movement in different contexts, such as, for example, for explaining organismal movement in movement ecology (Nathan et al. 2008), discussed in the related work Sect. 2.4.

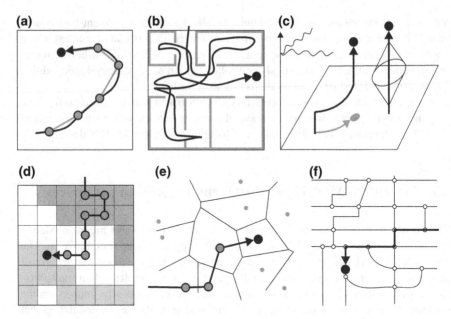

Fig. 2.1 Six basic space models accommodating the movement of point objects. **a** Euclidean homogeneous space, **b** constrained Euclidean space, **c** space-time aquarium, **d** heterogeneous field space, **e** irregular tessellation, **f** network space (**P9**. Laube 2009) (Reprinted from *Behaviour Monitoring and Interpretation, BMI, Smart Environments*, Gottfried, B. and Aghajan, H. (eds.), Laube, P., Progress in Movement Pattern Analysis, p. 49, Copyright (2009), with permission from IOS Press)

furthermore made apparent three dimensions discriminating conceptual movement spaces. These dimensions capture how movement is perceived from a physics perspective (Sect. 2.2.1), varying degrees of freedom of the moving objects in the movement spaces (Sect. 2.2.2), and the distinction between continuous and discrete spaces (Sect. 2.2.3).

2.2.1 Lagrangian Versus Eulerian Movement

Movement can be perceived from two different perspectives (see Fig. 2.2). The *Lagrangian* view considers changes in a moving object's location (**P19**. Both et al. 2013). This results in a stream of location fixes, typically in the form of (x, y, t)-tuples, describing movement as a two-dimensional, time-stamped polyline. The nodes represent the fixes, the straight-line edges between the nodes a simple approximation about the path taken in between. GPS tracking results in trajectories akin to the Lagrangian view. The *Eulerian* perspective describes movement as changes in location of moving objects relative to known, fixed points in space (**P19**. Both et al. 2013). Movement is perceived as a flux of objects passing by beacons, RFID tag

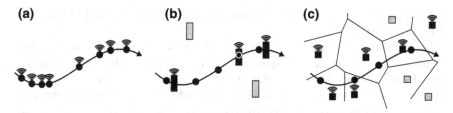

Fig. 2.2 Lagrangian versus Eulerian perspectives of movement. **a** The Lagrangian perspective focuses on the changes of location of the moving object, for example a GPS-tracked animal. The Eulerian perspective tracks moving objects as passing by fixed observations points, e.g., traffic gantries in **b** or GSM cells in **c**

readers, checkpoints or through gates or traffic gantries. In all these systems, the location of the checkpoints are known and fixed, and movement is captured in the form of the passing IDs and times of the MPOs. Recent developments in GSM and mobile ICT promote the later perspective, as more and more systems track moving objects in a checkpoint way.

Most research covered in this book adheres to the Lagrangian perspective. These studies have in common that a limited set of MPOs were tracked and their movement patterns analyzed. Examples studied in this book are racing pigeons for avian navigation research (**P3**. Laube et al. 2007), cows for precision farming (**P13**. Laube and Purves 2011), or fleet management issues with couriers (**P14**. Dodge et al. 2012). Equipping individuals with GPS receivers is ideal for studies where the test subjects are known and accessible, but the movement range is potentially unknown in advance.

By contrast, in some contexts the movement of individuals is constrained or bound to a limited number of channels or checkpoints (see Sect. 2.2.2). Here, the Eulerian perspective may have advantages as the individuals must eventually pass a checkpoint or gantry. For example, Both et al. (**P19**. 2013) and Bleisch et al. (**P20**. 2014) are based on a scenario for river health monitoring where fish are tracked via implanted RF transmitters when passing riverside RF readers in a simple topological river network. A system adhering to the Eulerian perspective requires less sophisticated equipment on the MPO side (RFID tags instead of GPS receivers and/or transmitters), and hence results in lighter tracking devices. The fish tracking example showed that the constraints imposed by the Eulerian perspective could be exploited for information gain in a decentralized data analysis scenario (see Chap. 4). Since fish eventually passed checkpoints when moving, algorithms running at the checkpoints were enabled to collect, enrich and exchange information about fish flows in the system as a whole.

One might argue that the Lagrangian perspective with a stream of GPS fixes offers a more precise tracking approach than the Eulerian perspective with its checkpoints with a potentially wide spacing. However, it should be noted that GPS data always is error prone and uncertain and the checkpoints location can be surveyed to very high precisions and at very fine spatial granularities. Hence, both perspectives can offer

tracking data with high precision, admittedly with a different notion of precision
(**P19**. Both et al. 2013).

Finally, a study reported on in Merki and Laube (**P16**. 2012) explicitly investigated
the influence of the choice of perspective when detecting interaction movement patterns. Here, simulated movement of interacting animals and observed movement of
children in an outdoor game were modeled and tracked using both perspectives and
respective conceptual data models. The comparative experiments revealed that the
different conceptual models of space required different formalizations of the patterns
to be detected. For example, adhering to the Eulerian perspective resulted in locational information of MPOs being discretized: The only locational information about
an MPO at any time was the current edge between adjacent checkpoint nodes. The
interaction pattern investigated in this study thus required a formalized neighborhood. For the Lagrangian case with its GPS fixes this neighborhood was modeled as
a disc with a given radius r. In the Eulerian case MPOs were formalized as neighboring when located on adjacent edges. These different formalizations not surprisingly
resulted in different analysis outcomes. Similarly, whereas the reaction time between
two objects meeting and one shying away proved crucial for the Lagrangian perspective, this criterion was less useful in the Eulerian perspective where again the given
edges dictated a coarse temporal granularity (see also the constraints issue discussed
in the next Sect. 2.2.2).

2.2.2 Constraints to Movement

A second important characteristic refers to the degree of freedom moving objects
have in their movement. Whereas some objects can (seemingly) move wherever they
wish (e.g., flying birds), others are limited where and how they can move across
space (e.g., pedestrians in an urban street network).

The assumption of unconstrained and hence free movement in an Euclidean space
is popular as it allows for a very simple conceptual model of space. However, this sim-
plistic assumption may ignore crucial constraints of the moving objects and hence
inadequately model their movement. Consider for instance animals in movement
ecology. Some animals can't swim which turns waterways into insurmountable
barriers, while others will not cross steep mountain ranges or need waterways to
swim in (**P20**. Bleisch et al. 2014). Hence, their seemingly free movement capacity is not free at all. Similarly, the homing pigeons studied in Laube et al. (**P3**.
2007) could indeed fly wherever they wanted, but the experimental setup was that
of a release-fly-back-to-loft scenario. Hence, movement azimuth distributions followed to a certain degree the given release site-loft configuration. The freedom of
movement can also be a matter of scale (see also Sect. 2.3.2). For instance, the
cross-scale movement analysis study in Laube and Purves (**P13**. 2011) revealed that
cows forage freely within their fenced paddock, but far reaching movement is limited by the fence. Whereas large scale foraging properties were not influenced by the
fence, edge effects became visible when investigating turning angles of the move-

ment trajectories at smaller scales, as the animals moved along the fence or even reversed direction when reaching the fence.

Most human movement in contrast is bound to some form of transportation infrastructure. This can be hiking paths, railway lines or motorways, or an inner city urban street network. The level of abstraction can hereby be varied depending on the aimed for analysis scale. As was shown in Richter et al. (**P18**. 2012) or Laube et al. (**P11**. 2010), an urban street network can be modeled as a simple graph consisting of nodes and edges. Depending on the data source, raw GPS data can then in a preprocessing step be map-matched to the closest edge. The same movement investigated at a finer spatial granularity, will require modeling the same urban street network as a configuration of polygons for streets and squares (see, for example, the simulated trajectories in Merki and Laube (**P16**. 2012)). Both et al. (**P19**. 2013) or Bleisch et al. (**P20**. 2014) illustrated that also animal movement can be constrained by a network movement space.

The choice of using an unconstrained or a constrained movement space has implications on the subsequent CMA process. For example, when mining movement patterns, the way a certain movement behavior can or can't be formalized using geometric constellations heavily depends on the underlying conceptual movement space. The leadership pattern introduced in Andersson et al. (**P5**. 2008) is based on the notion of a *front region*. Following the assumption of an unconstrained movement space, this front region was modeled as a wedge of edge length r and apex angle α, oriented in the current movement direction. For a study investigating similar interaction patterns (e.g., *pursuit and escape*), Merki and Laube (**P16**. 2012) found that the use of such a front region has limitations when using a constrained network space. With MPOs moving on typically straight edges, the effect of a front region becomes scale dependent. For objects on the same edge, a front region makes little sense, and for objects on adjacent edges, the configuration of the edges constrains relative positions of the involved objects.

This second dimension discriminating conceptual movement spaces (Sect. 2.2.2) has obvious links to the first dimension addressing the perspective taken on movement (Sect. 2.2.1). Movement models adhering to the Eulerian perspective often require constrained spaces: MPOs often move on edges between networked checkpoints or are allocated to Voronoi polygons of GSM cells. Similarly, both discussed dimensions are related to a third dimension addressing the issue of discretization of space. As a further level of abstraction or constraint, moving objects can be modeled as stepping between discrete space units, as will be discussed in the following section.

2.2.3 Continuous Versus Discrete Movement Spaces

Finally, the categorization in Fig. 2.1 revealed fundamental differences in using continuous or discrete movement spaces (top vs. bottom row in Fig. 2.1). Whereas in many cases the need for using discrete space models emerges from the data source, for example, when tracking mobile phone users through a sequence of visited GSM cells or Bluetooth beacons (Versichele et al. 2012), a discrete space can also be a deliberate design choice.

Proposing an approach for compressing GPS tracking logs, Richter et al. (**P18**. 2012) exploit the fact that urban transit is bound to a transportation network. First, instead of storing raw and highly redundant GPS logs, movement is abstracted to a time-stamped passage through a urban transit link (e.g., "along tram line #3" from stop #402 to #405 from 10:32 to 10:45). Second, following similar concepts introduced in the wayfinding literature, consecutive semantically equal passages are chunked together (e.g., a sequence of visited street edges is chunked to "along Bismarckstrasse" plus additional specifications). Here, the use of the discrete (and also constrained for that matter), semantically annotated transportation network allowed for the development of a compression technique for movement data. Similarly, movement events of fish from one river zone into another one are discrete movement events (**P20**. Bleisch et al. 2014).

The choice of a certain conceptual movement space can also allow for the adaption of related methods from neighboring research fields. For example, the abstraction of movement to a series of visited discrete places allows for the adoption of sequence and time series analysis. Du Mouza and Rigaux (2005) propose sequence queries for trajectories represented as sequences of visited GSM cells. Similar techniques are used in Dodge et al. (**P14**. 2012), however in that study not the movement space is discrete, but the trajectory is discretized in a segmentation process for similarity analysis (see Chap. 3).

2.3 Computing Movement Descriptors

Depending on the data capture procedure and the respective conceptual movement model, raw movement data comes as a stream of location data in the form of lists of GPS fixes or as time stamped visits to checkpoints. Apart from mapping movement traces for exploratory analysis, computing descriptive statistics capturing the essence of the studied movement process is a frequent entry point to CMA. Many tracking systems produce parameters describing the observed movement, such as instant speed, accelerometer readings, bearing and signal strength. Such system-produced data can carry useful information, however, the algorithmic basis of its computation is all too often unclear or even undocumented by the producer (**P13**. Laube and Purves 2011). The research summarized in this book suggests that maximal control and hence transparency is achieved when the movement descriptors underpinning CMA are (re-)computed from raw locational data.

2.3.1 Trajectory Operators

Laube et al. (**P3**. 2007) proposed the notion of trajectory operators, then called *lifeline context operators*, adopting Tomlin's map algebra for two-dimensional field data (Tomlin 1990) for the case of one-dimensional streams of movement fixes. A set

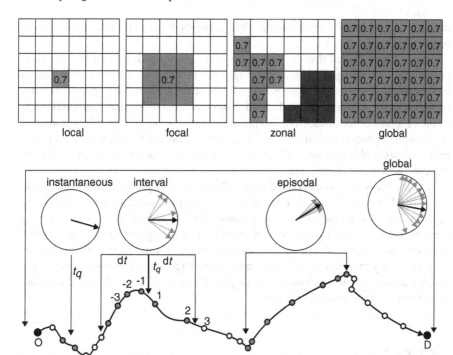

Fig. 2.3 Lifeline context operators in analogy to Tomlin's map algebra. Intervals can be delimited through a temporal window of fixed time (δt) or fixed number of fixes (± 3), adapted from Laube et al. (**P3**. 2007) (Reprinted from *Computers, Environment and Urban Systems*, 31(5), Laube, P., Dennis, T., Forer, P., and Walker, M., Movement Beyond the Snapshot—Dynamic Analysis of Geospatial Lifelines, page 486, Copyright (2007), with permission from Elsevier)

of instantaneous (relating to Tomlin's "local" operators), interval ("focal"), episodal ("zonal") and global ("global") operators were suggested for computing descriptive movement parameters (see Fig. 2.3). Interval operators, just as the focal operations in the two-dimensional case, compute movement descriptors at any fix along a trajectory as a function of the temporal fix neighborhood. This fix neighborhood can be defined through a defined number of neighboring fixes, or, allowing for irregular sampling or missing values, through a defined interval. Also similar to the 2D case, weighted neighborhood functions were discussed, assigning temporally close fixes higher weights.

That study not only featured the above listed commonly used movement descriptors but also exemplified the development of additional measures tailored towards a specific application field, here avian navigation research. *Navigational displacement* measures at any given point along the trajectory the deviation angle of a homing bird from the direct path to its loft. *Approaching rate* measures whether or not and to what degree a bird moves towards loft. These two measures were developed in a close and iterative collaboration with the application scientists, tailored towards their specific needs.

That study indicated that the seemingly straightforward derivation of movement properties (such as speed, azimuth, turning angle, or sinuosity) allows for substantial methodological diversity. There is not just one way of computing speed, azimuth, or sinuosity for movement data. Instead, from the various possible combinations of (i) data capture procedures, (ii) conceptual movement models, (iii) different notions of movement properties found in different application fields, and finally, (iv) variable analysis scales (see Sect. 2.3.2), results a surprising diversity of approaches to compute movement descriptors.

In many ways the work in Laube et al. (**P3**. 2007) was a predecessor to later work, especially Laube and Purves (**P13**. 2011). The outlook section of the 2007 study suggested that the notion of interval operators, there conceptualized as a smoothing operator for imperfect trajectory data, would allow for systematically varying the analysis scale, aiming at investigating movement data at variable temporal granularities. The study reported on in Laube and Purves (**P13**. 2011) followed that idea and presented experiments systematically varying the temporal granularity of deriving movement descriptors from trajectories.

2.3.2 Scale

Scale is a quintessential geographic concept. All three meanings of scale—cartographic scale, analysis scale, and phenomenon scale (Montello 2001)—are relevant to CMA.

Cartographic scale expresses the relationship between the earth's surface and its necessarily much smaller depiction on a map (Montello 2001). This is clearly relevant as the visual display of movement traces is an entry point to CMA. However, GIScience and related application sciences have so far given little attention to methodological challenges around cartographic scale of mapping movement, which is surprising given typically large and heterogeneous raw data volumes and the discipline's rich history in aggregation and generalization. Trajectories are either mapped in their entirety as polylines (**P10**. Dennis et al. 2010) or aggregated to density maps for giving a quick overview (see, for example, Fig. 8 in **P3**. Laube et al. 2007). Aggregation and generalization of trajectories remains an important topic for further research.

Analysis scale refers to the granularity at which phenomena are measured or aggregated (Montello 2001). Whenever movement is modeled as trajectories, analysis scale refers to the spacing of the fixes, that is the spatial and/or temporal separation of location measurements along the movement trace. Laube and Purves (**P13**. 2011) made the point that the granularity of data capture (the inbuilt or user-set sampling rate of the tracking system) does not necessarily prejudice the subsequent analysis scale. By contrast, the experimental piece investigated the implications of varying the temporal analysis scale at which the movement descriptors speed, turning angle

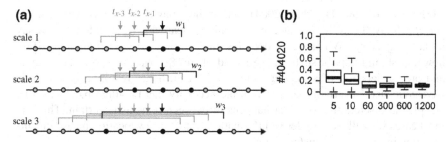

Fig. 2.4 Cross-scale derivation of movement descriptors. **a** Systematic variation of the interval operator width w between *black sampling points* when computing, for example, speed. **b** Dropping speed values for cow #404020 with coarser sampling rates (in ms^{-1}). Adapted from Laube and Purves (**P13**. 2011) (Republished from Laube, P. and Purves, R., How fast is a cow? Cross-scale Analysis of Movement Data, *Transactions in GIS*, 15(3), pp. 401–418, 2011, John Wiley & Sons Ltd, DOI:10.1111/j.1467-9671.2011.01256.x.)

and sinuosity were derived. Methodologically the study illustrates the adaptation of methods to CMA that proved useful in other geo-disciplines. The methods design draws on analogies to the classic Fisher et al. (2004) multi-scale piece "Where is Helvellyn?" that showed how the computation of slope or the labeling of landforms may vary with the sampling point spacing, that is the analysis scale.

Given tracking data of cows with a fine sampling rate of a fix every 0.25 s, the study derived movement descriptors for six temporal scales w = [5 s, 10 s, 1 min, 5 min, 10 min, 30 min] (Fig. 2.4a). The study indeed found that the results for various movement descriptors vary with the used analysis scale. For example, it confirmed earlier findings, for example mentioned in Laube et al. (**P3**. 2007), that speed values drop with coarser sampling, as the straight-line connectors between wide spaced fixes systematically underestimate the actual path travelled (Fig. 2.4b). However, it also became obvious that such cross-scale effects should not be discussed without a careful consideration of the uncertainty of the original GPS data (see Sect. 2.3.3).

Finally, *phenomenon scale* refers to the region over which geographic processes (here movement) occur (Montello 2001). The intuitive rule requires that the analysis scale matches the actual phenomenon scale. However, just as other geographic phenomena, it is often not a priori evident what that phenomenon scale is, or movement processes can even express characteristics at and across different scales. Think of a penguin leaving fine grained movement traces on a ice sheet floating on a continental ocean current. Methods for up-scaling and down-scaling knowledge remain an open challenge in CMA.

2.3.3 Uncertainty and Data Quality

Uncertainty is another classic theme in GIScience. Uncertainty is an unavoidable property of the world, information about the world and our cognition about the world (Worboys and Duckham 2004). Uncertainty may arise because of *uncertain specifications*. It may, for instance, not be entirely clear what we refer to when we say a "daily trajectory". When exactly does this start and end? Do we exclude stops? If so, how long should stops be such that we exclude them? Second, measuring the accurate and precise location of a moving object is difficult, hence resulting in *uncertain measurements*. Uncertainty arises from the necessity that moving objects must be sampled at discrete times—what happens in between remains uncertain. Third, in most cases we will want to derive information from our raw location measurements resulting in *uncertain transformations*.

Laube and Purves (**P13**. 2011) showed that such uncertainty of GPS data should not be neglected, especially when investigating movement at fine spatio-temporal granularities. During that study initially aiming to discover multi-scale effects when computing movement descriptors, it became obvious that what was assumed to be "raw" GPS data in the first place was indeed smoothed by algorithmic post-processing, adding positional uncertainty to the fixes. The study consequently extended its focus and developed a methodological framework for giving an indication of those temporal scales for which the influence of uncertainty was less important than the actual signal, the characteristics of the trajectory. To that end Monte Carlo Simulation was used to model the uncertainty of the fixes. Each fix was assigned an uncertainty sampled from a bivariate standard deviation before the movement parameters were recalculated. T-tests then indicated at what scales and with what uncertainties the found distributions for original and MC-simulated descriptors were significantly different. This methodology—another GIScience classic—revealed for the example studied in the paper, that for an uncertainty of 1 m speed is reliably computed for temporal scales of 60 s and greater.

Imfeld et al. (**P1**. 2006) illustrate the importance of quality control of raw data in a field experiment. The study goes at great length to get an idea of the positional accuracy of movement and movement context data. The field experiment also showed that specifically point in polygon tests for linking fixes to the environment are sensitive to both inaccuracies in the location data but also the context information. This article indicated that the implications of inaccurate data may depend on the task at hand. If the goal is aggregation (for example the production of a density map) then positional error is not such an issue. However, when point in polygon links are used, the influence of the error may be larger, depending on the analysis scale.

2.4 Related Work

This section summarizes selected related work complementing and completing the discussion of the topics covered in this chapter. The chapter then concludes with insights and lessons learned from both the research covered in this chapter and from the related work.

Conceptual models for movement and movement spaces. Nathan et al. (2008) present an often cited conceptual framework aiming at a unifying generic theory in the application field of movement ecology. The framework is based on four inter-acting mechanistic components of organismal movement: 1. the internal state (why do animals move?), 2. motion capacity (how do they move?), 3. navigation capacity (when and where to move?), and 4. the external factors affecting movement. The primary goal of their conceptual framework is providing a basis for hypothesis gen-eration and contributing to a better understanding of the causes of movement and its role in ecological and evolutionary processes. Hence, whereas Nathan et al. (2008) aims at understanding movement as a process in ecology, the work summarized in this book rather aims at conceptual models required for abstracting and representing movement in spatio-temporal information systems.

Having said that, I agree with Nathan et al. (2008) that CMA requires a firm integration of the movement paths and the embedding movement space. By contrast, much related work in GIScience and computer science mainly focuses on the move-ment path or trajectory alone. However, the following definitions illustrate the crucial role of the underlying movement space, even though in most work this interrelation is not explicitly discussed.

Work focusing on the shape and arrangement of movement traces is typically based on some definition of a *trajectory* as a polyline in a 2D Euclidean space that can self-intersect, built by a set of time stamped points (Gudmundsson et al. 2007). This geometry-oriented perspective is useful for geometry-based operations, as is for example illustrated in Gudmundsson et al. (2009) using a variant of the Douglas–Peucker path-simplification algorithm for compressing large volumes of movement data (similar to the motivation in Richter et al. (**P18**. 2012) but using a rather different approach). The importance of the embedding movement space becomes obvious when movement is modeled as the path an object moved along in a transportation network. For example, Cao and Wolfson (2005) define a *road-snapped trajectory* as a set of visited edges in a 2D transportation network. Similarly, Kuijpers et al. (2010) model network-based trajectories in a 3D space-time prism. In contrast to such merely geometric definitions, for others a trajectory has a semantic loading. Early work on temporal GIS for spatio-temporal reasoning about people's personal lifelines anticipated the importance of semantic enrichment of event histories (Thériault et al. 2002). In their piece "on a conceptual view of trajectories" Spaccapietra et al. (2008, p.130) define a trajectory as a "user defined record of the evolution of the position of an object that is moving in space during a given time interval in order to achieve a given goal". Some others go even further and require that raw movement data is first cleaned and preprocessed, even interpolated and segmented before a trajectory in a narrower sense is created (Yan et al. 2008). The semantic perspective has also produced a large amount of research on formal and qualitative modeling of movement trajectories. For instance, (Noyon et al. 2005) propose a *spatio-temporal trajectory* (STT) abstract data type and related operations suitable for representing and querying semantic-based trajectories.

Whereas the last decade has seen a constant stream of work studying *Lagrangian*, GPS-based movement data, recent years have seen the arrival of more and more work benefitting from the increasing availability of *Eulerian* movement data (from GSM mobile phone systems, RFID, WiFi, Bluetooth). Even though not explicitly referencing the Lagrange and Euler dichotomy, Andrienko et al. (2008) present a similar categorization of various ways of observing movement. They discuss *time-based recordings* (regular interval sampling), *change-based recording* (record made when position changes), *location-based recording* (when an object passes a beacon, Eulerian perspective), *event-based recording* (position fix made with phone call), and combinations of the above.

Records of mobile phone companies provide the most prominent source for Eulerian movement data, although rigorous and justifiable privacy concerns limit their availability. Typically, such data is also tied to a constrained movement space (street network) and discrete. The work by the group around Rein Ahas on mobile phone usage in the small European country Estonia may serve as a demonstrative example for this line of work (Ahas et al. 2009, 2010; Silm and Ahas 2010). The group has access to passive mobile positioning data of a large fraction of the Estonian population. *Passive* refers to the fact that instead of actively requesting records of a moving object, here only times, anonymized caller ID, and cell ID with the geographical coordinates of the antenna are recorded when a user happens to make a phone call or connect to the internet. Nevertheless, with an average of approximately six fixed calls per user and day, such Eulerian movement paths can be constructed as sequences of visited antennas. Exploiting this rather extensive coverage of a small country's population, the group produces interesting work related to modeling home and work locations (Ahas et al. 2009, 2010) or short-term population mobility (Silm and Ahas 2010).

Whereas many Lagrangian/GPS studies are based on experimental set-ups where tracking devices are distributed and then monitored, many Eulerian studies piggyback on existing ICT infrastructure, e.g., people's individual phones and the GSM (Global System for Mobile Communications) networks maintained by mobile phone companies. As the Estonian example shows, such secondary exploitation has the potential for accessing much larger numbers of individuals. For example, Versichele et al. (2012) use proximity-based Bluetooth tracking at a large festival estimating flow maps of up to 10 % of the festival's 1.5 million visitors. Delafontaine et al. (2012) demonstrate with a similar Bluetooth setup the use of sequence alignment methods for revealing variability in the visiting patterns at a trade fair, with patterns formalizing the number and order of visited exhibition halls.

Computing movement descriptors. Even though the various application fields study movement with a rather diverse range of motivations, there seems to be a limited set of parameters characterizing movement. For example, Andrienko et al. (2008) list movement-related characteristics around *position, direction, speed, change of direction, change of speed, acceleration,* and *travel distance.* Similarly, Dodge et al. (2008) list *primitive parameters* (e.g., position (x, y)), *primary derivatives* (e.g., speed $f(x, y, t)$), and *secondary derivatives* (e.g., acceleration $f(speed)$). Hence, whereas most authors agree on a basic set of parameters, it is much more ambiguous how these are to be computed, and the implications of such choices when the parameters serve as the fundamental ingredients of any subsequent analysis. A data challenge based on GPS tracking of lesser black-backed gulls illustrated in 2011 a rather impressive variation when several experts in movement analysis were asked to compute the same basic movement parameters (trip distance and duration, mean speed; Shamoun-Baranes et al. 2012). So, whereas there is still surprisingly little work on the implications of choices of conceptual models and data structures, issues related to data quality and granularity are often investigated, especially in application-driven movement analysis research fields such as ecology or transportation research, where careful conduct with data emerges from a long tradition of empirical research.

Scale. Whereas data-driven and theoretical research areas tend to simply accept a given spatial or temporal granularity, in problem-driven research areas, most prominently in movement ecology, it is widely acknowledged that the observed movement signal varies with different observation and analysis scales (Nams 2005; Fryxell et al. 2008). Since often the data representing the different scales derived from rather different data capture procedures (as for example Fryxell et al. 2008, in VHS, GPS and tracks in the fresh snow), it remains difficult to separate multi-scale effects of used methods from differences of the data capturing techniques. Some exceptions explicitly performed multi-scale analysis like that of Laube and Purves (**P13**. 2011) aiming at isolating methodological effects. For example, Nams (2005) derives fractal dimension D (as a measure for sinuosity, or tortuosity, as it is often termed in behavioral ecology) for the same trajectories at various spatial scales. The hypothesis underlying this research states that animals express different movement behaviors (e.g., tortuous foraging at fine spatial scales but directed advances at coarse spatial scales) at different spatial scale sections (so called "domains"), which are identified through cross-scale analysis. Similar work studied the influence of the sampling regime on the computation of the home ranges (Borger et al. 2006) and the computation of a straightness index (Postlethwaite et al. 2013). Such work is important but remains difficult since obtaining the required fine spatio-temporal granularities still is difficult and costly.

Uncertainty. It is widely accepted that uncertainty is an inherent property of movement data (Andrienko et al. 2008). Giannotti and Pedreschi (2008) list measurement error and unavoidable discrete sampling regimes as two major sources of imperfection for movement data. It is however interesting to observe that only some areas concerned with movement analysis address this uncertainty while others prefer to largely neglect it. Most theoretical work studying trajectories mainly as geometric features defines in their preliminaries that fix locations are perfectly known and accurate (for example, Gudmundsson et al. 2007; Benkert et al. 2008). However, it is known that especially in urban areas GPS can be inaccurate, having an effect on the actual analysis task. For example, imperfection of tracking data had implications for the segmentation and travel mode allocation in NYC (Gong et al. 2012) and for the size of movement pattern clusters in pedestrian movement in a recreational application (Moreira et al. 2010). Whereas such methodological research remains rare in the GIScience and computational geometry fields, application areas with urgent applied research questions such as for example, again, movement ecology, have a much stronger interest in the implications of imperfect data on the actual outcomes of the analytical process. For instance, Hurford (2009) shows in an experimental piece the emergence of systematic bias when computing turning angle due to GPS measurement error. Similarly, Jerde and Visscher (2005) use Monte Carlo simulation to quantify the measurement error for estimates of turning angle and step length as a function of distance between consecutive locations. By contrast, in the database community, imperfect tracking data has for some time been the driving force for a significant research strand around the handling and querying of uncertain positional data in moving object databases (MOD, for example, Trajcevski et al. 2004).

2.5 Concluding Remarks

This chapter has summarized contributions from a series of articles underpinning the methodological fundamentals of computational movement analysis—conceptual modeling and abstraction, and representation and description of movement spaces as well as the moving entities embedded therein. It was shown that from a growing diversity of technologies allowing for the tracking of individuals emerges a wide range of different forms of movement data, adhering to an equally diverse range of conceptual models and data structures. Movement of individuals is captured from GPS, WiFi, Bluetooth, cell phone logs, ticketing and intelligent public transit cards, as well as gantry stations. Most research analyzing movement is very problem driven, even data driven, and often the infrastructure setting dictates how the world is abstracted, and hence the conceptual data models are chosen. In other words, the way data is collected often rules the abstraction of the world, which then has implications on the analysis process. However, since there is little comparison between methods, there is little insight about the implications of such crucial design choices. CMA studies how the diversity of how we perceive and model movement has implications on how we describe and quantify movement, and hence how we progress in the CMA process enriching movement data to process knowledge, as will be studied in the following chapter on movement mining.

Computational movement analysis contributes to the theory of GIScience by adapting and adopting core concepts of spatio-temporal modeling and analysis to movement data as a relatively new form of geographic information. Work was portrayed adapting 2D field operators to 1D streams of location fixes, both based on a deeply geographic notion of spatial respectively temporal dependence or neighborhood. Other studies borrowed from the methodological toolbox of geomorphometry and performed multi-scale analysis addressing sampling issues and Monte Carlo simulation investigating uncertainty.

Movement data often inherits its conceptual model and sampling regime from a given tracking system or research design from the application scientist collecting the data in the first place. However, I argue that computational movement analysis can and must move beyond accepting such preliminaries as unchangeable constraints and rather consider them as design choices and systematically study the implications of such design choices. Furthermore, as became evident, for example, in the work on semantic trajectory compression, the characteristics of conceptual movement spaces can at the same time be a limitation but also an opportunity. This will become even more evident in the following two chapters.

References

Ahas, R., Silm, S., Järv, O., Saluveer, E., & Tiru, M. (2010). Using mobile positioning data to model locations meaningful to users of mobile phones. *Journal of Urban Technology, 17*(1), 3–27.

Ahas, R., Silm, S., Saluveer, E., & Järv, O. (2009). Modelling home and work locations of populations using passive mobile positioning data. In G. Gartner & K. Rehrl (Eds.), *Location based services and telecartography II* (pp. 301–315)., Lecture Notes in Geoinformation and Cartography Berlin: Springer.

Andersson, M., Gudmundsson, J., Laube, P., & Wolle, T. (2008). Reporting leaders and followers among trajectories of moving point objects. *GeoInformatica, 12*(4), 497–528.

Andrienko, N., Andrienko, G., Pelekis, N., & Spaccapietra, S. (2008). Basic concepts of movement data. In F. Giannotti & D. Pedreschi (Eds.), *Mobility, data mining and privacy* (pp. 15–38). Berlin: Springer.

Benkert, M., Gudmundsson, J., Hübner, F., & Wolle, T. (2008). Reporting flock patterns. *Computational Geometry, 41*(3), 111–125.

Bleisch, S., Duckham, M., Galton, A., Laube, P., & Lyon, J. (2014). Mining candidate causal relationships in movement patterns. *International Journal of Geographical Information Science, 28*(2), 363–382.

Borger, L., Franconi, N., De Michele, G., Gantz, A., Meschi, F., Manica, A., et al. (2006). Effects of sampling regime on the mean and variance of home range size estimates. *Journal of Animal Ecology, 75*(6), 1393–1405.

Both, A., Duckham, M., Laube, P., Wark, T., & Yeoman, J. (2013). Decentralized monitoring of moving objects in a transportation network augmented with checkpoints. *The Computer Journal, 56*(12), 1432–1449.

Cao, H., & Wolfson, O. (2005). Nonmaterialized motion information in transport networks. In T. Eiter & L. Libkin (Eds.), *Database theory—ICDT 2005, proceedings* (Vol. 3363, pp. 173–188)., Lecture Notes in Computer Science Berlin: Springer.

Delafontaine, M., Versichele, M., Neutens, T., & Van de Weghe, N. (2012). Analysing spatiotemporal sequences in bluetooth tracking data. *Applied Geography, 34*, 659–668.

Dennis, T. E., Chen, W. C., Koefoed, I. M., Lacoursiere, C. J., Walker, M. M., Laube, P., et al. (2010). Performance characteristics of small global-positioning-system tracking collars for terrestrial animals. *Wildlife Biology in Practice, 6*(1), 14–31.

Dodge, S., Laube, P., & Weibel, R. (2012). Movement similarity assessment using symbolic representation of trajectories. *International Journal of Geographical Information Science, 26*(9), 1563–1588.

Dodge, S., Weibel, R., & Lautenschütz, A.-K. (2008). Towards a taxonomy of movement patterns. *Information Visualization, 7*(3–4), 240–252.

Du Mouza, C., & Rigaux, P. (2005). Mobility patterns. *GeoInformatica, 9*(4), 297–319.

Fisher, P., Wood, J., & Cheng, T. (2004). Where is Helvellyn? Fuzziness of multi-scale landscape morphometry. *Transactions of the Institute of British Geographers, 29*(1), 106–128.

Fryxell, J. M., Hazell, M., Börger, L., Dalziel, B. D., Haydon, D. T., Morales, J. M., et al. (2008). Multiple movement modes by large herbivores at multiple spatiotemporal scales. *Proceedings of the National Academy of Sciences, 105*(49), 19114–19119.

Giannotti, F., & Pedreschi, D. (2008). Mobility, data mining and privacy: A vision of convergence. In F. Giannotti & D. Pedreschi (Eds.), *Mobility, data mining and privacy* (pp. 1–11). Berlin: Springer.

Gong, H., Chen, C., Bialostozky, E., & Lawson, C. T. (2012). A GPS/GIS method for travel mode detection in New York City. *Computers, Environment and Urban Systems, 36*(2), 131–139.

Gudmundsson, J., Katajainen, J., Merrick, D., Ong, C., & Wolle, T. (2009). Compressing spatiotemporal trajectories. *Computational Geometry, 42*(9), 825–841.

Gudmundsson, J., van Kreveld, M., & Speckmann, B. (2007). Efficient detection of patterns in 2D trajectories of moving points. *GeoInformatica, 11*(2), 195–215.

Hägerstrand, T. (1970). What about people in regional science. *Papers of the Regional Science Association*, *24*, 7–21.

Hurford, A. (2009). GPS measurement error gives rise to spurious 180°-turning angles and strong directional biases in animal movement data. *PLoS ONE*, *4*(5), e5632.

Imfeld, S., Haller, R., & Laube, P. (2006). Positional accuracy of biological research data in GIS— A case study in the swiss national park. In M. Caetano & M. Painho (Eds.), *7th International Symposium on Spatial Accuracy Assessment in Natural Resources and Environmental Sciences* (pp. 275–280). Portugal: Lisbon.

Jerde, C. L., & Visscher, D. R. (2005). GPS measurement error influences on movement model parameterization. *Ecological Applications*, *15*(3), 806–810.

Kuijpers, B., Miller, H. J., Neutens, T., & Othman, W. (2010). Anchor uncertainty and space-time prisms on road networks. *International Journal of Geographical Information Science*, *24*(8), 1223–1248.

Laube, P. (2009). Progress in movement pattern analysis. In B. Gottfried & H. Aghajan (Eds.), *Behaviour Monitoring and Interpretation, BMI, Smart Environments* (Vol. 3, pp. 43–71)., Ambient Intelligence and Smart Environments Amsterdam, NL: IOS Press.

Laube, P., & Dennis, T. (2006). Exploratory analysis of movement trajectories. In GeoCart,. (2006). *National Cartographic Conference*. Auckland, NZ.

Laube, P., Dennis, T., Walker, M., & Forer, P. (2007). Movement beyond the snapshot - dynamic analysis of geospatial lifelines. *Computers, Environment and Urban Systems*, *31*(5), 481–501.

Laube, P., Duckham, M., & Palaniswami, M. (2011). Deferred decentralized movement pattern mining for geosensor networks. *International Journal of Geographical Information Science*, *25*(2), 273–292.

Laube, P., Duckham, M., & Wolle, T. (2008). Decentralized movement pattern detection amongst mobile geosensor nodes. In T. J. Cova, K. Beard, M. F. Goodchild, & A. U. Frank (Eds.), *GIScience 2008* (Vol. 5266, pp. 199–216)., Lecture Notes in Computer Science, Springer: Berlin Heidelberg.

Laube, P., Duckham, M., Worboys, M., & Joyce, T. (2010). Decentralized spatial computing in urban environments. In B. Jiang & X. Yao (Eds.), *Geospatial analysis and modelling of urban structure and dynamics, geojournal library* (pp. 53–74). Berlin: Springer.

Laube, P., & Purves, R. S. (2011). How fast is a cow? Cross-scale analysis of movement data. *Transactions in GIS*, *15*(3), 401–418.

Merki, M., & Laube, P. (2012). Detecting reaction movement patterns in trajectory data. In J. Gensel, D. Josselin, & D. Vandenbroucke (Eds.), *AGILE'2012 International Conference on Geographic Information Science*. FR: Avignon.

Montello, D. (2001). Scale in geography. In N. J. Smelser & P. B. Baltes (Eds.), *International Encyclopedia of the social and behavioral sciences* (pp. 3501–13504). Oxford: Pergamon Press.

Moreira, A., Santos, M. Y., Wachowicz, M., & Orellana, D. (2010). The impact of data quality in the context of pedestrian movement analysis. In M. Painho, M. Y. Santos, & H. Pundt (Eds.), *Geospatial thinking* (pp. 61–78)., Lecture Notes in Geoinformation and Cartography Berlin: Springer.

Nams, V. O. (2005). Using animal movement paths to measure response to spatial scale. *Oecologia*, *143*(2), 179–188.

Nathan, R., Getz, W. M., Revilla, E., Holyoak, M., Kadmon, R., Saltz, D., et al. (2008). A movement ecology paradigm for unifying organismal movement research. *Proceedings of the National Academy of Sciences*, *105*(49), 19052–19059.

Noyon, V., Devogele, T., & Claramunt, C. (2005). A formal model for representing point trajectories in two-dimensional spaces. In J. Akoka, S. Liddle, I.-Y. Song, M. Bertolotto, I. Comyn-Wattiau, W.-J. Heuvel, M. Kolp, J. Trujillo, C. Kop, & H. Mayr (Eds.), *Perspectives in conceptual modeling* (Vol. 3770, pp. 208–217)., Lecture Notes in Computer Science Berlin: Springer.

Peuquet, D. J. (2002). *Representation of space and time*. London, UK: The Guilford Press.

Postlethwaite, C. M., Brown, P., & Dennis, T. E. (2013). A new multi-scale measure for analysing animal movement data. *Journal of Theoretical Biology*, *317*, 175–185.

Richter, K.-F., Schmid, F., & Laube, P. (2012). Semantic trajectory compression: Representing urban movement in a nutshell. *JOSIS*, *4*, 3–30.

Shamoun-Baranes, J., van Loon, E. E., Purves, R. S., Speckmann, B., Weiskopf, D., & Camphuysen, C. J. (2012). Analysis and visualization of animal movement. *Biology Letters*, *8*(1), 6–9.

Silm, S., & Ahas, R. (2010). The seasonal variability of population in estonian municipalities. *Environment and Planning A*, *42*(10), 2527–2546.

Spaccapietra, S., Parent, C., Damiani, M. L., de Macedo, J. A., Portoa, F., & Vangenot, C. (2008). A conceptual view on trajectories. *Data and Knowledge Engineering*, *65*(1), 126–146.

Thériault, M., Claramunt, C., Séguin, A. M., & Villeneuve, P. (2002). Temporal GIS and statistical modelling of personal lifelines. In D. E. Richardson & P. van Oosterom (Eds.), *Advances in Geographic Information Systems Research II: Proceedings ot the Interantional Symposium on Spatial Data Handling, Delft* (pp. 433–449). Berlin: Springer.

Tomlin, C. D. (1990). *Geographic information systems and cartographic modeling*. Englewood Cliffs: Prentice Hall.

Trajcevski, G., Wolfson, O., Hinrichs, K., & Chamberlain, S. (2004). Managing uncertainty in moving objects databases. *ACM Transactions on Database Systems (TODS)*, *29*(3), 463–507.

Versichele, M., Neutens, T., Delafontaine, M., & Van de Weghe, N. (2012). The use of bluetooth for analysing spatiotemporal dynamics of human movement at mass events: A case study of the ghent festivities. *Applied Geography*, *32*(2), 208–220.

Worboys, M., & Duckham, M. (2004). *GIS—A computing perspective* (2nd ed.). New York: CRC Press.

Yan, Z., Macedo, J., Parent, C., & Spaccapietra, S. (2008). Trajectory ontologies and queries. *Transactions in GIS*, *12*, 75–91.

Chapter 3
Movement Mining

With ever increasing volumes and complexity of spatio-temporal information, knowledge discovery in databases and its best known step data mining, have rapidly gained importance within Geography and GIScience. Analyzing spatio-temporal data first of all means structuring data, then extracting relevant spatial patterns and rules and providing decision makers with enriched information and condensed knowledge rather than flooding them with raw data. Movement patterns, for example, represent such sought-for high-level process knowledge derived from low-level trajectory data. This second chapter introducing the research field of Computational Movement Analysis (CMA) reviews research on several aspects of mining movement data, including the conceptualization and formalization of movement patterns and the development of algorithms for their detection, the computing of trajectory similarity, and methods for visualization-based exploratory analysis of movement data.[1]

Overarching research objectives. The research summarized in this chapter contributes to the following overarching research objectives of computational movement analysis.

- Illustrate to what respect the conceptual underpinning and toolset of data mining suits the specific requirements of computational movement analysis.
- Exemplify how geographically-informed movement mining contributes to a better understanding of movement processes.
- Promote a more thorough attitude towards evaluation of proposed movement mining methods.

---

[1] Whereas this chapter discusses movement mining in conventional omniscient and centralized information systems or databases, the following Chap. 4 discusses the rather peculiar case where data mining is performed in decentralized systems such as geosensor networks. Even though most of the work summarized in Chap. 4 nominally also proposes data mining techniques, its theoretical underpinning in decentralized spatial computing justifies a separate chapter focusing on decentralized movement analysis alone.

© The Author(s) 2014
P. Laube, *Computational Movement Analysis*,
SpringerBriefs in Computer Science, DOI 10.1007/978-3-319-10268-9_3

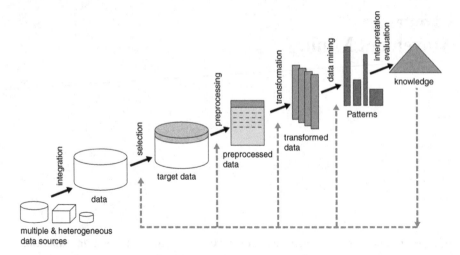

Fig. 3.1 The KDD process and its core step data mining, adapted from Fayyad et al. (1996)

3.1 Data Mining for CMA

According to Fayyad et al. (1996, p. 39) data mining refers to one particular step in the overall process of discovering useful knowledge in data (Fig. 3.1).

> *Data mining* is the application of specific algorithms for extracting patterns from data. [...] The additional steps in the knowledge discovery in databases (KDD) process, such as data preparation, data selection, data cleaning, incorporation of appropriate prior knowledge, and proper interpretation of the results, are essential to ensure that useful knowledge is derived from the data. Blind application of data-mining methods (rightly criticized as data dredging in the statistical literature) can be a dangerous activity, easily leading to the discovery of meaningless and invalid patterns.

3.1.1 Defining Movement Mining

Miller and Han (2009, p. 3) build on such early definitions of KDD and data mining when they outline the fundamentals of Geographic Data Mining and Knowledge Discovery. In their words, data mining involves distilling data into information or facts about the domain described by a database. KDD by contrast is then the higher-level process enriching such found information or facts into knowledge through interpretation of information and its integration with existing knowledge about the domain. This notion of distilling data into information and further into knowledge complies very much with Anthony Galton's call to bridge the gulf between low level observational data and the high-level conceptual schemes in which humans think and understand geographic phenomena (Galton 2005, p. 300). The key building blocks of this distillation process leading from data through information to knowledge are *interesting patterns*. These are introduced as non-random properties and relationships that are valid, novel, useful, and ultimately understandable (Fayyad et al. 1996; Miller and Han 2009).

It is this notion of data mining that builds the theoretical underpinning of the notion of movement mining used here:

Definition *Movement mining* aims for conceptualizing and detecting non-random properties and relationships in movement data that are valid, novel, useful, and ultimately understandable.

Even though the definition of data mining implies large data sources, the core elements of the definition refer to qualities rather than quantities. Instead of defining movement mining through particular techniques or methods such as artificial intelligence, machine learning, statistics, or database systems, this book adheres to a conceptual view of qualifying the outcomes of the analytical process. The movement mining process aims for the ideal of finding properties and relationships, in a wider sense, any form of structure in the data, patterns or trends, segmentations, similarities, or clusters, that measure up to the given qualities.

The qualities *valid, novel, useful,* and *ultimately understandable,* in accordance to Fayyad et al. (1996) and Miller and Han (2009, p. 3), are in the following illustrated for the special case of movement mining using the example of the movement pattern leadership (**P5**. Andersson et al. 2008). *Leadership* here is defined as the situation when in a group of moving entities "one object is leading others", in the sense that this object spatially leads the way and the others follow for some time (Fig. 3.2).

- *valid*—properties and relationships should be general enough to apply to new data, hence they should not just capture an anomaly or a peculiarity of the current data. Although initially inspired by coordination in gray wolves (Peterson et al. 2002) or grazing heifers (Dumont et al. 2005), patterns describing collective motion pattern such as *leadership* or *flock* (Laube et al. 2005;

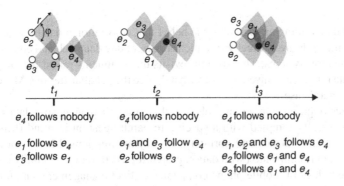

Fig. 3.2 Movement pattern example *leadership*, Andersson et al. (**P5**. 2008)

Benkert et al. 2008) are also investigated for pedestrians or tourists (Orellana and Renso 2010). A valid movement pattern should hence be sufficiently generic and abstract such that it can potentially be found in a diverse range of application domains.

- *novel*—properties and relationships should be nontrivial and unexpected. The leadership pattern requires a specified set of moving objects to maintain a specific topological relation for a given temporal interval. This involves complex coordination and arrangement amongst the moving objects. Contrasting with the leadership definition in Andersson at al. (**P5**. 2008), Nagy et al. (2010) reveal the rather unexpected insight that for flocking birds the leader actually guides the flock from a position in the back rather than in front of the flock. The quality "novel" emphasizes data mining's claim to detect the unexpected, rather than being confirmatory.
- *ultimately understandable*—properties and relationships should be simple and interpretable by humans, for example, domain experts. Even though the use of terms that may have variable and potentially ambiguous semantic connotations has raised criticism, the pictorial language of patterns exemplified through the pattern *leadership* has very much added to the appeal and accessibility of movement patterns, especially when collaborating with domain experts in movement ecology or urban transit.
- *useful*—properties and relationships should lead to some effective action, e.g. a successful decision making or scientific investigation. The pivotal challenge in data mining and also movement mining is the identification of the useful patterns in the plethora of potential patterns. To this end, Laube and Purves (2006) compared in a Monte Carlo approach movement patterns found in real data with patterns found in synthetic data generated trough a form of random walk. More recently community activities targeting specifically a comparison of analytical approaches have contributed to the discussion about the usefulness of found patterns and trends (Shamoun-Baranes et al. 2012).

3.1.2 What is Special About Movement Data?

KDD and data mining are based on the belief that conventional database query and statistical methods are not really suitable for current day data-rich and computation-rich information systems. This section revisits the arguments raised for the general case of data mining and investigates to what degree the peculiarities of CMA comply with the general case.

Statistics require clean and noiseless, rather small or manageable data sets that were scientifically sampled with a specific research question in mind (Miller and Han 2009). Independence and normality are key assumptions. Movement mining, however, often is confronted with data captured for a different, potentially unrelated purpose (e.g. logs of mobile phone companies or fleet management data for buses, taxis or rental bikes). Furthermore, tracking data is highly spatio-temporally autocorrelated (**P18**. Richter et al. 2012). Hence, just as in data mining in general, movement mining is challenged by noisy and uncertain, multi-source data.

Miller and Han (2009) furthermore stress that data mining and KDD is more inductive than traditional deductive statistical analysis. Statistics aim to confirm *a priori* formulated hypotheses, based on some theory. By contrast, the patterns and relations hidden in large data sources sought in a data mining and KDD process are by definition unexpected and unknown in advance. At the least, it would be very difficult to have a complete *a priori* picture of what to find. Hence data mining is most useful when applied early in the process of scientific discovery, when structuring large data sources, aiming in an exploratory way towards the establishment of a theory. This way of scientific reasoning is very much prevalent in the movement mining literature, especially when combined with scientific visualization in visual analytics (Andrienko and Andrienko 2007).

Summarizing why data mining as a technique is especially suitable for CMA, Table 3.1 compares the peculiarities of movement data identified in Sect. 1.1 with the arguments motivating data mining or conventional statistics. Besides an arguably good general fit, the comparison also reveals commonalities regarding especially the type of the investigated data (noisy, uncertain) and the type of patterns of interest (non-trivial, unexpected complex relations).

3.2 Movement Mining Tasks

There are many different, yet similar, categorizations of data mining tasks (see for example Chakrabarti et al. 2006; Hand et al. 2001; Han and Kamber 2006, for an

Table 3.1 Comparison of the properties of movement data and the strengths of data mining as a tool, contrasted to the characteristics of conventional database querying and statistics

Peculiarities of movement data (Sect. 1.1)	Motivation for data mining	Motivation for conventional DB queries and confirmatory statistics
Geographic reference systems	–	–
Permanent change	–	–
Complex objects	Non-trivial relationships	Simple data points in spreadsheets
Implicit relations	Discover the unexpected	Confirmatory, confirm the expected
Overlap	Multi-source, ad hoc integrated data sources	Solitary spreadsheets
Spatial dependency and heterogeneity	Spatio-temporally autocorrelated	Independence, normality
Uncertainty	Noisy and uncertain data	Clean, noiseless
Derivative data	Multi-source, ad hoc integrated data sources	Data sampled for primary scientific question
Scale issues	–	–

overview). Loosely adhering to the structures given there, the movement mining
work included in this section is grouped into four subsections.

- Segmentation and filtering
- Similarity and clustering
- Movement patterns
- Exploratory analysis and visualization

Figure 3.3 gives an overview of movement mining tasks. The iconic examples are
embedded in the six space models accommodating the movement of point objects
introduced in Fig. 2.1 in Sect. 2. The following section exemplifies the development
and application of movement mining techniques through original research included
in this book.

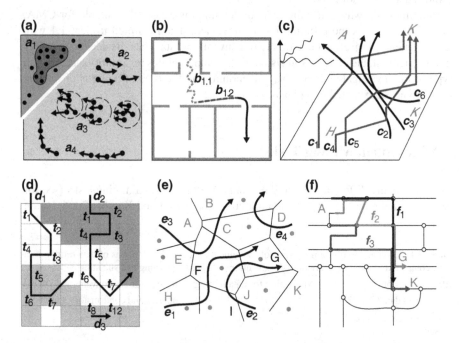

Fig. 3.3 Examples for movement mining tasks embedded in typical conceptual movement spaces. **a**
Exploratory analysis and movement patterns, *home range* (a_1) and arrangement patterns *leadership*
(a_2), *flock* (a_3), *single file* (a_4); **b** segmentation into segments expressing different sinuosity ($b_{1.1}$,
$b_{1.2}$); **c** similarity and clustering: similar 3D shapes or origin-destination (from H to K vs. from K
to A). **d** similarity and clustering: d_1 and d_2 show highly synchronous movement, d_3 is an outlier;
e sequence patterns: two trajectories both featuring a sequence I, F, G; **f** similarity and clustering
in a network space; f_1 and f_2 are more similar than f_1 and f_3, all three trajectories build an origin-
destination cluster (from A to K) (**P9**. Laube 2009) (Reprinted from *Behaviour Monitoring and
Interpretation, BMI, Smart Environments*, Gottfried, B. and Aghajan, H. (eds.), Laube, P., Progress
in Movement Pattern Analysis, p. 55, Copyright (2009), with permission from IOS Press)

3.2.1 Segmentation and Filtering

Most tracking systems adopting the Lagrangian perspective of movement (such as GPS or VHF receivers) produce streams of location fixes, irrespective of the current movement behavior of the tracked object. Here, segmentation is a useful approach for structuring large volumes of raw movement data. Similar to image segmentation in image processing (Shapiro and Stockman 2001), *segmentation* then refers to the process of partitioning movement data into multiple segments, with the goal of simplifying or changing the representation of the trajectory into something that is more meaningful or easier to analyze. Buchin et al. (2011b) specifically define segmentation as partitioning a trajectory into a (typically small) number of pieces, where the obtained segments have uniform characteristics.

In the movement mining process segmentation and filtering can take the function of preprocessing steps, aiming at reducing noise and condensing the signal for a given analytical task (Fig. 3.1). First the trajectory is segmented into coherent segments, then only those segments relevant to the analysis task are selected for subsequent processing. In the most basic case this involves separating stops from moves. Many movement data sets contain large periods when the object is not moving. For instance, the GPS trackers producing the data in Laube and Purves (**P13.** 2011) used a movement model based on a Kalman filter, where even when the object is immobile, a smooth curve is fitted to the location fixes resulting in trajectory coils (see Fig. 3.4). Since such "pseudo movement" interfered with the multi-scale study

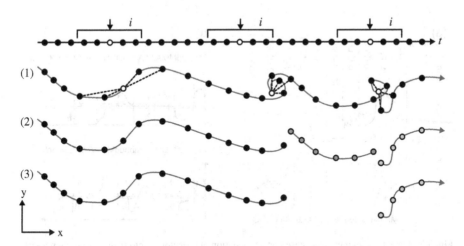

Fig. 3.4 Trajectory segmentation and filtering. (*1*) Application of an interval operator for computing the average Euclidean distance to other fixes inside a temporal window *i*. (*2*) Removal of all points where average distance is less than a given threshold, i.e., filtering of static points, (*3*) filtering of subtrajectories with less than a threshold temporal length. Adapted from Laube and Purves (**P13.** 2011) (Republished from Laube, P. and Purves, R., How fast is a cow? Cross-scale Analysis of Movement Data, *Transactions in GIS*, 15(3), pp. 401–418, 2011, John Wiley & Sons Ltd, DOI: 10.1111/j.1467-9671.2011.01256.x)

it had to be removed through a segmentation and filtering step. Similar filtering was also required in Laube et al. (**P12**. 2011a) working with the same cattle tracking data. Many segmentation approaches are based on the idea that in an immobile phase the area covered by the object during that interval must be smaller than when the object is moving. Laube et al. (**P12**. 2011a) use a minimal enclosing circle, whereas Laube and Purves (**P13**. 2011) based their stop detection on an average Euclidean distance to other fixes inside a temporal window i to be less than some threshold d (see Fig. 3.4).

Furthermore, segmentation can be based on the shape of the trajectory. Trajectories can be split into segments of similar straightness or sinuosity, or at sharp turns. Alternatively trajectory segmenting can be performed based on any descriptive parameter assigned to individual fixes. A stop could then simply be identified by, for instance, low speed values (**P14**. Dodge et al. 2012). Either parameters such as speed, acceleration, heading, or sinuosity emerge from the primary sensor system or they are derived through instantaneous or interval trajectory operators (see **P3**. Laube et al. 2007, Sect. 2.3.1). Segments or subtrajectories are then delineated based on sequences of uniform parameters. Dodge et al. (**P14**. 2012) go one step further and categorize the derived movement parameters (MP) into a set of predefined classes, creating a string-like representation of the characteristic of the trajectory (see Fig. 3.5). These MP class sequences are then used for assessing the similarity between "strings", or trajectories respectively (see Sect. 3.2.2).

Whereas most examples in this chapter adhere to a preprocessing notion of segmentation, it can also be the primary movement mining task of a study. Applications

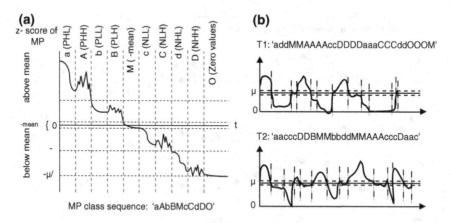

Fig. 3.5 Trajectory segmentation based on a translation of trajectories into string-like sequences of movement parameter (MP) classes. MP classes capture the deviation from the mean of a parameter and the sinuosity of the MP profile, for example a(PHL) reads as "Positive High (deviation from mean) and Low (sinuosity). Adapted from Dodge et al. (**P14**. 2012) (Republished from Dodge, S., Laube, P., and Weibel, R., Movement Similarity Assessment Using Symbolic Representation of Trajectories. *International Journal of Geographical Information Science*, 26(9), pp. 1563–1588, 2012, Taylor & Francis, DOI:10.1080/13658816.2011.630003)

include behavior classification in animal ecology (stand, forage, fly) or travel mode detection (walk, bike, car, bus; see Sect. 3.4).

3.2.2 Similarity and Clustering

Trajectories and other traces of moving objects are complex objects and therefore comparing trajectories in order to assess their similarity or for clustering is a challenging problem. Trajectories, for example, can significantly vary in length or extent, shape and orientation, as well as granularity. Furthermore, there are different notions of what might be considered similar when comparing trajectories. Trajectories could be considered similar if they have similar shapes (elongated vs. clumped), share commonalities in terms of derived movement parameters (similar average speed or sinuosity), visit similar places (edge G in Fig. 3.3f), feature similar patterns (such as repeated bursts of relocation in Levy walks, Gonzalez et al. 2008) or express sequences (Fig. 3.3e) or diurnal rhythms. Depending on the application domain the notion of similarity will focus on different aspects of the complex spatio-temporal traces of moving objects.

Here, a general procedure for assessing trajectory similarity and subsequent trajectory clustering is presented and illustrated through the respective procedure in Dodge et al. (**P14**. 2012).

- *Specify a frame of reference.* What shall be compared? The entire lifeline of an object? Yearly migrations of animals or daily commuting trips of people? To this end, segmentation methods may be used in preprocessing steps. Dodge et al. (**P14**. 2012) investigate the (a) whole lifelines of hurricanes from the moment of their formation until their degradation, and (b) the movement along a specifically selected set of edges of an urban transportation network.
- *Choose or define a distance metric.* Dodge et al. (**P14**. 2012) specifically argue for a spatio-temporal notion of distance, explicitly going beyond only considering the atemporal geometry. The trajectory is thereto transformed into a sequence of class labels based on the movement parameter speed. Then a modified edit distance for comparing such strings is used as a distance metric.
- *Compute similarity matrix and apply a suitable clustering technique.* To this end, Dodge et al. (**P14**. 2012) applied complete-linkage agglomerative hierarchical clustering.

Figure 3.6 illustrates the procedure for four selected hurricane trajectories. Even though H2 and H3 appear to have rather similar shapes, in terms of speed sequences H1 and H2 express the smallest distances and then, the largest similarity, respectively.

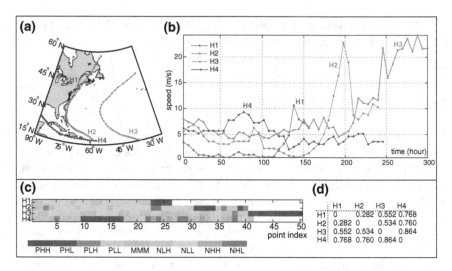

Fig. 3.6 Trajectory similarity. Computation of normalized weighted edit distance (NWED) for four hurricane trajectories. **a** spatial footprint of hurricane trajectories; **b** hurricane speed profiles; **c** segmented speed profiles; **d** NWED pair-wise distance matrix (**P14**. Dodge et al. 2012) (Republished from Dodge, S., Laube, P., and Weibel, R., Movement Similarity Assessment Using Symbolic Representation of Trajectories. *International Journal of Geographical Information Science*, 26(9), pp. 1563–1588, 2012, Taylor & Francis, DOI:10.1080/13658816.2011.630003)

3.2.3 Movement Patterns

Mining movement patterns is the quintessence of movement mining. This book features work on coordination patterns such as *leadership* in Andersson et al. (**P5**. 2008) and *flocking*[2] in Laube et al. (**P6**. 2008b) and Laube et al. (**P12**. 2011a), and Both et al. (**P19**. 2013) as well as reaction patterns such as *pursuit and escape*, *confrontation*, or *avoidance* in Merki and Laube (**P16**. 2012). Examples for movement patterns are illustrated for *leadership* in Fig. 3.2 and *pursuit and escape* in Fig. 3.7. This section revisits the work on movement patterns included in this book, reviews strengths and weaknesses of the included work and thereon presents three suggestions for good practice in movement mining.

3.2.3.1 Definitions Grounded in Application Theory

Definitions of movement patterns should be grounded in the theory of the respective application domain. Andersson et al. (**P15**. 2008) base the conceptualization of the pattern *leadership* on detailed descriptions of the involved processes and

---

[2] Note, the research on *flocking* featured in this book combines data mining concepts with decentralized spatial computing principles. This chapter focuses on the data general data mining aspects, Chap. 4 on the specifics of mining movement patterns in a decentralized setting.

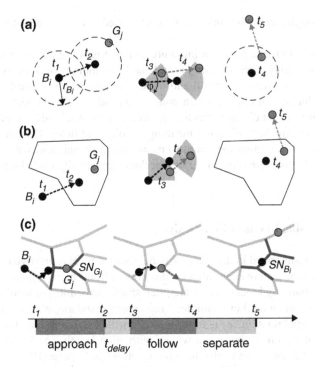

Fig. 3.7 Pursuit and escape pattern (**P16**. Merki and Laube 2012). An actor (*black dot*) approaches a reactor (*grey dot*), they follow each other (front region φ), and finally they separate again. **a** Lagrangian perspective, **b** Eulerian perspective, **c** Eulerian perspective for network bound movement (Republished from Merki, M. and Laube, P., Detecting reaction movement patterns in trajectory data, In Gensel, J., Josselin, D., and Vandenbroucke, D. (eds.), *Bridging the Geographic Information Sciences*, International AGILE2012 Conference, Avignon (France), April, 24–27, Lecture Notes in Geoinformation and Cartography, Springer, Berlin, Germany, ISBN 978-3-642-29063-3, Copyright © 2012)

spatio-temporal constellations found in various publications in the movement ecology and animation industry bodies of literature. These sources provide precise descriptions of what it means to "initiate the collective movement of a coordinated group" of grazing heifers or "leading a pack of wolves". Similarly, Merki and Laube (**P16**. 2012) developed their formal notion of interaction patterns based on work on predator/prey behavior and territorial interaction found in the ecology literature as well as studies on suspicious behavior of pedestrians in video surveillance.

Early work on movement patterns (Laube et al. 2005) was rather tools-driven, where pattern definitions were heavily influenced by the algorithms at hand for the subsequent data mining process. Whereas these papers were important for promoting the idea of movement patterns in the first place, gained insights about the limitations of such definitions lead to more problem-driven definitions. For example, Laube et al. (**P12**. 2011a) revealed the limitation of a disc-shaped notion of proximity for the pattern flock. For a group of grazing cows, group coherence is better expressed through piece-wise linkage, as is often used in convoy definitions (Jeung et al. 2008a, b).

Borrowing from the application literature can be a blessing and a curse when it comes to terminology. Clearly, using established terminology and respective definitions helps conceptualizing useful and understandable patterns (see Sect. 3.1.1). However, problems can emerge when conceptually identical patterns have different yet similar terms in different fields (e.g., flocks, convoys, herds), or even worse when identical terms have differing meanings in different fields. For that reason it can be observed that terminology in movement pattern mining has evolved towards more structural terms (moving cluster) rather than semantically loaded terms (flock, convoy).

3.2.3.2 Increasing Levels of Complexity

The research summarized in this chapter indicates that novel (non-trivial and unexpected) movement patterns can often be decomposed into more primitive building blocks. These building blocks in turn are often spatial proximity- or topology-based relations or set relations. For example, *leadership* in Andersson et al. (**P5**. 2008) requires the spatial relation "e_j is in front of e_i", which is in turn based on the concept of a front region $front(e_i)$, a wedge with edge length r and apex angle α. This very same front region is also featured in Merki and Laube (**P16**. 2012), here, however, acting as a building block of the interaction pattern *pursuit and escape* (see Fig. 3.7). The latter piece furthermore features very similar pattern primitives for "proximity" (even in two different ways for both the Lagrangian and the Eulerian perspective, see Fig. 3.7 and Sect. 2.2.1) and for "change of movement direction". The sequence patterns in Bleisch et al. (**P20**. 2014) are per definition chained-up simple movement or environmental events ("rapid upstream movement", "moderate water temperature"), building more complex sequences. Such a sequence could, for example, comprise of a moderate water temperature event, followed by two rapid upstream movement events ($\{wt3_e\}, \{u_e\}, \{u_e\}$).

The hierarchical decomposition of movement patterns furthermore assists the development of algorithms for detecting the patterns. Andersson et al. (**P5**. 2008) first introduce a series of auxiliary data structures in the form of precomputed data arrays that are then later combined for an efficient detection of the patterns. These arrays store for each moving entity the number of consecutive unit time intervals expressing a certain follow-behavior. For example, leadership can be derived from the arrays "the number of time units e_j has at least m followers" and "the number of time units e_j is not following anyone else". The algorithmic procedure in Merki and Laube (**P16**. 2012) computes in a very similar way first all primitive events contributing to a pattern and then investigates the required sequences (see Fig. 3.7, *approach*, followed by *follow* after a delay, followed by *separate*).

The usefulness of such a hierarchical composition of movement patterns is further emphasized through several attempts of a categorization or ontology of movement patterns. Dodge et al. (2008) refer to *primitive patterns* and *compound patterns*,

Andrienko and Andrienko (2007) stipulate the combination of *individual movement behaviors* (IMB) to form *dynamic collective behaviors* (DCB). Irrespective of the precise nature of the building blocks, there seems to be an agreement that higher-level behavior patterns emerge when lower-level building blocks are strung together in a temporally ordered sequence. Merki and Laube (**P16**. 2012) even showed that not only the sequence matters, but that the temporal spacing between lined-up building blocks plays a crucial role when separating related yet different behaviors.

3.2.3.3 Non-deterministic Pattern Mining

Even though movement pattern mining qualifies as a "retrieval by content" data mining task (for example "detect all leaders leading m followers for k time units"), its true strength lies in exploratory rather than confirmatory analysis, in hypothesis forming rather than hypothesis testing. Movement patterns are a good example of how movement mining is strongest when prompting researchers to new and unexpected relationships in data. The exact extent of a leadership pattern is not very relevant. The knowledge that in a certain data set there exist leadership patterns in the first place, and the order of magnitude of such patterns, is much more relevant.

Similarly, in most cases there is not *a priori* knowledge about the precise extent or any parameter or threshold specifying a movement pattern. How many followers must an alpha wolf have and for how long must they follow it? In some cases the domain specific literature may give indications about parameters specifying movement patterns. However, it is the repeatability of the algorithmic search for such patterns that allows for series of sensitivity experiments, lowering the influence of potentially arbitrarily chosen thresholds in the knowledge discovery process. Merki and Laube (**P16**. 2012), for example, assessed the sensitivity of the parameters ϕ, the angle of the front region for *pursuit and escape*, and delay d between an approach and a separation in a *confrontation* pattern. Laube et al. (**P12**. 2011a) explicitly studied the sensitivity of the data mining process with respect to the chosen size of flocking patterns (here with grazing cows). Performing movement mining in such an exploratory way turns a potential weakness into a strength as it allows exactly the required embedding of the somewhat mechanistic pattern mining in further domain expertise, through a close collaboration of data mining experts with domain experts (Fayyad et al. 1996, p. 39).

3.2.4 Exploratory Analysis and Visualization

Given its specific characteristics, movement data presents an ideal use case for spatio-temporal exploratory analysis, visualization, and visual analytics concepts (Andrienko et al. 2010). Irrespective of the precise label, the core idea is to combine the strengths of human and computational data processing (Keim et al. 2008). Just as is outlined in the previous sections, algorithmic techniques are used to prepare and condense the raw movement data by suggesting similarities and clusters or segments and patterns. The structured data is then visualized and the human analyst can use his/her strength of confirming the expected and detecting the unexpected by visually inspecting the displayed information, often in an interactive analysis setting (Thomas and Cook 2006; Andrienko and Andrienko 2007).

Even though the focus of the work summarized in this book is clearly neither visualization nor visual analytics, the section on movement mining tasks concludes with selected examples where visual and exploratory approaches support the wider movement mining process. These examples are:

- *Small multiples.* Small multiples (Tufte and Graves-Morris 1983) or collections (Bertin et al. 1981) are a set of juxtaposed data representations allowing sliced insight into multivariate data. Given that movement data often involves large data sets of multiple objects with overlapping and repetitive space use, small multiples are a widespread tool in movement mining. The first example included here features small multiples for nine maps illustrating the space-use patterns of nine GPS-tagged common brushtail possums (**P10**. Dennis et al. 2010, Fig. 3, p. 23). The second example illustrates for ten individual cows three different movement parameters at six temporal scales each (**P13**. Laube and Purves 2011, Fig. 6, p. 412).
- *Aggregation.* Laube et al. (**P3**. 2007, Fig. 13, p. 495) features an example of aggregation. Here, the spatially explicit property of flight sinuosity of a large set of trajectories was interpolated into one field aggregating the information of all pigeons. Figure 3.8 reveals that the homing pigeons show highest sinuosities around the release site, indicating a phase of re-orientation after release.
- *Interactive interfaces.* Even though not explicitly being a result of the research summarized in this volume, several interactive interfaces were developed and intensively used in the movement mining process, especially for plausibility testing. For example, the agent-based simulation environment REPAST served as a base for an interactive data mining interface for the development of decentralized flock mining algorithms in Laube et al. (**P6**. 2008b) and Laube et al. (**P12**. 2011a). The interfaces allowed for the live interactive adjustment of several algorithm parameters during simulation runs, with linked windows showing the effects of these adjustments in a map view as well as error plots (see Fig. 3.9).

3.3 Evaluation

As pointed out above, the unreflected application of data mining methods can easily lead to the discovery of meaningless patterns (Fayyad et al. 1996). Similar care is required for the development of data mining techniques, and hence also movement mining techniques. Evaluating if the developed methods are sound and produce useful and meaningful knowledge is at the same time very important and very difficult. This section discusses concepts that can help evaluating the quality of proposed movement mining techniques. *Verification, validation,* and *credibility* are based on the terminology for evaluating and testing ecological models by Rykiel (1996), whilst *efficiency* reflects basic considerations of computing costs and applicability for practical use.

- *Verification* is the technical matter that relates to how accurately the analytical ideas are translated into computer code or mathematical formalisms. Verification aims at assuring that a method (a.k.a. data mining technique) produces mechanically and logically correct results. For example, debugging code is a verification process. In terms of algorithm analysis verification is related to *algorithmic efficiency*. Sometimes verification is referred to as an internal evaluation.

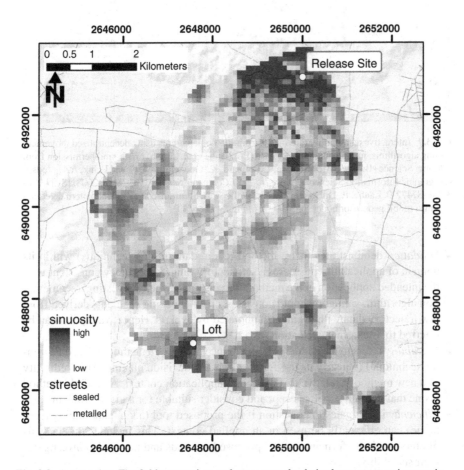

Fig. 3.8 Aggregation. The field summarizes and aggregates the derived parameter *trajectory sinuosity* for 54 racing pigeon in a field experiment. The aggregation reveals highest sinuosity values around the release site (**P3**. Laube et al. 2007) (Reprinted from *Computers, Environment and Urban Systems*, 31(5), Laube, P., Dennis, T., Forer, P., and Walker, M., Movement Beyond the Snapshot—Dynamic Analysis of Geospatial Lifelines, p. 495, Copyright (2007), with permission from Elsevier)

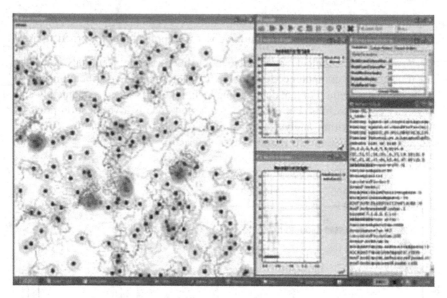

Fig. 3.9 Interactive data mining interface for mining flock patterns using decentralized movement mining algorithms, implemented in REPAST (**P6**. Laube et al. 2008b) (With kind permission from Springer Science+Business Media: Cova, T. J., Beard, K., Goodchild, M. F., and Frank, A. U., (eds.), Geographic Information Science, *Lecture Notes in Computer Science*, vol. 5266, 2008, ISBN 978-3-540-87472-0, Laube, P., Duckham, M., and Wolle, T., Decentralised movement pattern detection amongst mobil geosensor nodes, pp. 199–216, Fig. 5)

- *Validation* demonstrates that a method (a.k.a. data mining technique) within its domain of applicability possesses a satisfactory range of accuracy consistent with the intended application of the method. For instance, when suggesting a segmentation algorithm, validation could consist of cross-checking segments derived from raw trajectory data with segments annotated by experts. Hence, validation can be referred to as external evaluation.
- *Credibility* refers to the degree of belief in the validity of a method (a.k.a. data mining technique) to justify its use for research and decision making. The credibility of a new method is relative to a particular application context. A credible method is one that domain experts accept and consider suitable for a given problem.
- *Efficiency* investigates the question if the proposed tool (a.k.a. data mining technique) can efficiently be used in an applied context. This includes the analysis of algorithmic performance of the proposed methods and hence also investigates their scalability.

Note the similarities to the qualities of patterns outlined in Sect. 3.1.1. The work summarized in this book repeatedly addresses the above concepts. In the following, selected examples are singled out illustrating the concepts in the general context of movement mining in computational movement analysis. Verification is a basic imperative when developing CMA methods and is hence rarely evident from scientific publications. For that reason it is not further investigated in the following.

3.3.1 Validation

Validation investigates if and how well a new method does what it is supposed to do. Often this involves a comparison of method outcomes with respective observations or manual measurements made by experts. Rykiel (1996) lists a wide range of different validation tests. In the context of this chapter five shall be portrayed in more detail.

3.3.1.1 Face Validity

For assessing the face validity of a method knowledgeable people are asked if the method and its behavior is reasonable, if input-output relationships appear reasonable. For instance, discussions with farming experts when working on Laube et al. (**P12**. 2011a) revealed that the tool-driven definition of the movement pattern *flock* was not optimal. The pattern definition used required the individuals to move within a circular disc of a given radius, whereas the observed movement rather showed flocks as chains of piecewise connected pairs. Here, the face validity test revealed a limited suitability of the chosen formalization underlying the proposed method.

3.3.1.2 Visualization Techniques

Another validation strategy is offered by visualization, exploratory analysis, or visual analytics, where the data mining process is combined with a human analyst. Here the user directly inspects the method outcomes by the use of visual displays and thereby validates the plausibility of method outcomes. Examples for visual validation can be found in Sect. 3.2.4.

3.3.1.3 Internal Validity

For assessing the internal validity of a method test data sets can be used for investigating if the method produces and reproduces a consistent output. The error analysis carried out for decentralized flock mining in Laube et al. (**P6**. 2008b) and Laube et al. (**P12**. 2011a) shall serve here as an example for testing the internal validity. Given a data set with a known spatio-temporal occurrence of target patterns, the number of actually present patterns is compared to the number of patterns found by the movement mining algorithm. *Error of omission* (eoo, "missed patterns") accounts for existing patterns not found, while *error of commission* (eoc, "false positives") specifies the wrongly detected patterns when no pattern actually exists.

Laube et al. (**P12**. 2011a) illustrated a typical dilemma for the assessment of internal validity. A major difficulty lies in finding suitable data sets that (a) express exactly the patterns to be detected, and (b) feature sufficient semantic information documenting those patterns. For example, the fine-grained cow tracking data used in Laube et al. (**P12**. 2011a) did not feature information about the spatially and temporally varying composition and arrangement of the tracked group of cows. Consequently, assumptions had to be made for the validation process that certainly are up to debate. This shortcoming can be overcome when synthetic data is generated where the number and distribution of patterns can precisely be controlled for experiments (as has been illustrated above). This, however, can undermine the credibility or the generic character of a proposed technique since one could argue that the simulation was inappropriately fitted to suit the data mining technique. As a code of conduct this book suggest to aim for a combination of both simulated and real data for both validation and verification purposes.

3.3.1.4 Sensitivity Analysis

Movement mining methods may require the setting of parameters, and consequently can express variable sensitivity with respect to these parameters. Sensitivity analysis investigates which parameters cause significant changes in the methods' outcomes. The core argument of Laube and Purves (**P13**. 2011) is based on a sensitivity analysis. The paper investigates the sensitivity of methods for computing movement descriptors to the selected analysis scale and associated data uncertainties. In this case the validation procedure highlighted crucial sensitivities that are often neglected.

Another form of sensitivity analysis is performed in the movement mining approaches presented in Laube et al. (**P6**. 2008b) and Laube et al. (**P12**. 2011a). Both studies required some form of algorithm parameterization, balancing eoo versus eoc in the constrained decentralized computing environment (see Sect. 4.1). Here, the varied constraint is the size of the communication range c relative to the pattern radius p: The larger the communication range, the smaller eoo and the larger eoc (Fig. 3.10). The figure also illustrates that specifically error of omission can be reduced when the rigor of the task is relaxed from finding complete flock patterns ($n = 10$ individuals) to "partial" flocks built of fewer individuals.

3.3.1.5 Comparison to Other Methods

When new methods extend other methods then a direct comparison of their outcomes is a suitable validation means. Dodge et al. (**P14**. 2012) propose a new variant of an edit distance for assessing trajectory similarity, that clearly is positioned in a succession of related methods. The comparative study then revealed similarities and differences between the compared methods, allowing a validation of the newly proposed method.

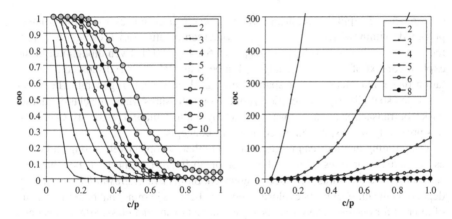

Fig. 3.10 Balancing *eoo* versus *eoc* in decentralized flock mining, simulated movement data (**P12**. Laube et al. 2011a) (Republished from Laube, P., Duckham, M., Palaniswami, M., Deferred Decentralized Movement Pattern Mining for Geosensor Networks, *International Journal of Geographical Information Science*, 25(2), pp. 273–292, 2011, Taylor & Francis, DOI:10.1080/13658810903296630)

3.3.2 Credibility

Method evaluation should also make sure that a method meets the needs of its users, assuring a certain notion of acceptance and suitability with domain experts. The work summarized in this book strongly supports the argument that data mining and hence movement mining methods must prove useful to users in an application context. Clearly, testing suggested methods with real data supports credibility. Real data emerging an application context was used in Merki and Laube (**P16**. 2012, tracked students in an outdoor game), Dodge et al. (**P14**. 2012, hurricanes, couriers), and Laube et al. (**P12**. 2011a, cows in a smart farming study). Earlier work by Laube and Purves (2006) investigated, for example, the relevance of movement patterns based on the notion of *interestingness*. I argue that the interestingness measures proposed by Silberschatz and Tuzhilin (1996) and Geng and Hamilton (2006) also build a useful starting point for assessing the credibility of patterns in movement mining.

Objective interestingness measures depend solely on the structure of the pattern and the underlying data. The most commonly known objective measure for the quality, strength or interestingness of data mining rules are support and confidence given for association rules. *Support* is generally defined as the frequency of a pattern in a data set, while *confidence* expresses the prediction strength of the rule (Mohammad and Nishida 2010). In Bleisch et al. (**P20**. 2014) support and confidence measures were adapted for mining candidate causal relationships between movement events and environmental states, such as "allows", "initiates" or "terminates". The study developed a sequence mining approach relating environmental states ("high water temperature", "high river flow") with movement events of fish moving in a river network ("upstream movement"). For example, in 84 occurrences of rapid upstream movement (um), 6 occurrences were immediately preceded (within two days) by a moderate water temperature (mwt) event, i.e., resulting in a $confidence_{event}(E_{um} \rightarrow E_{mwt}) = 6/84 = 0.071$. Even though not in the explicit context of movement analysis, Laube et al. (**P4**. 2008a) discussed an adaptation of similar measures for spatio-temporal data mining in general. The paper suggested spatially explicit definitions of the two measures where the classic market basket or transaction metaphor was replaced by spatio-temporal proximity. Crucially, this proximity was expressed based on fuzzy concepts.

Data mining methods can produce large numbers of objectively strong and interesting patterns or rules, that are however of no interest to the user. For that reason, subjective measures have been suggested. *Subjective interestingness measures* depend on the class of users exploring the data, bearing in mind that patterns that are of interest for one user class, may be of no interest to another class. Silberschatz and Tuzhilin (1996) identify two reasons why a pattern is interesting from a subjective point of view: *unexpectedness*, which indicates how surprising the pattern is to a user, and *actionability* which indicates whether the user can act on the pattern to his/her advantage.

3.3.3 Efficiency

Useful movement mining techniques should comply with minimal requirements in terms of computing costs. To this end, efficiency is typically used to describe properties of an algorithm relating to how much of various types of resources it consumes. Hence, the notion of efficiency can vary depending on which resource is of special interest. For example, in the limited computing environments of geosensor nodes, the number of messages required for completing a task can be a vital performance factor. In Both et al. (**P19**. 2013) mobile sensor nodes are tasked with monitoring the flow in a cordon-structured transportation network. Figure 3.11 then illustrates the scalability of three related proposed algorithms in terms of the number of messages sent when the number of fish (a.k.a. sensor nodes) in the system is increased. The three algorithms represent three levels of complexity of decentralized communication (1. wired cordons, 2. fish carry information packages in between cordons, 3. fish also exchange information packages as they meet). The figure confirms the expectation that algorithms 1 and 2 scale linearly, whilst algorithm 3 (added fish-fish communication) expresses in the worst case a communication complexity that scales with the square of the total number of fish F, hence $O(|F|^2)$.

This book includes a wide variety of papers, some having a more conceptual focus whilst others have a more algorithmic character. Some conceptual papers tolerate roughly quadratic running times, but indicate possible optimization strategies. For instance the string-matching process used in Dodge et al. (**P14**. 2012) is admittedly rather slow (i.e., $O(n^2)$ with n representing the number of "letters" in the trajectory). However, pruning techniques have been used for related problems and could be also be implemented for an operational use of the proposed concepts. By contrast,

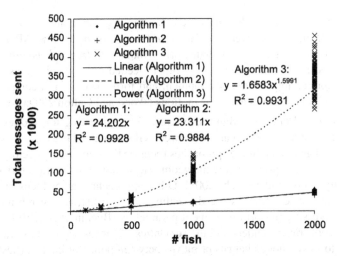

Fig. 3.11 Scalability of communication in terms of total number of messages sent with change in number of fish for Algorithms 1–3 (**P19**. Both et al. 2013) (Republished from Both, A., Duckham, M., Laube, P., Wark, T., and Yeoman, J., Decentralized monitoring of moving objects in a transportation network augmented with checkpoints, *The Computer Journal*, 2013, 56(12), pp. 1432–1449, DOI: 10.1093/comjnl/bxs117, by permission of Oxford University Press)

other work included in this book studies alternative versions of proposed algorithms, explicitly aiming at optimization of running times. For instance Andersson et al. (**P5**. 2008) proposed improved algorithms for leadership detection using additional index structures (buckets) where the running times strongly depend on the characteristics of the instances.

3.4 Related Work

This section summarizes further related work relevant to the topics covered in this chapter. The chapter concludes with insights and lessons learned from both research covered in this chapter and in related work.

Data mining for CMA. Data mining techniques are proposed and applied in a wide range of areas studying very diverse movement phenomena (**P15**. Laube et al. 2011b). Given the diversity of application problems and the multidisciplinary backgrounds of the involved scientists, it is perhaps little surprising that there seems to be slow progress towards an accepted general theory of movement mining. Nevertheless, there is work aiming towards general taxonomies or ontologies of movement patterns and movement mining tasks in general. Dodge et al. (2008) propose a taxonomy for movement patterns, distinguishing between *generic patterns* (e.g. co-occurrence, moving clusters) and *behavioral patterns* that are particular to certain types of moving objects (e.g. foraging or flocking birds). Andrienko and Andrienko (2007) first focus on patterns of individuals (i.e., changes in position and other movement characteristics of an entity over time, *individual movement behaviors*, IMB) that may then be found in groups of entities to form *dynamic collective behaviors* (DCB). Finally, Wood and Galton contribute a much needed ontological analysis of collective motion, aiming for the development and formalization of a comprehensive classification of collectives and their motions (Wood and Galton 2009a, b).

Segmentation and filtering. Many techniques for segmenting trajectories emerge from transdisciplinary collaborations between GIScience, databases, and especially computational geometry. Some approaches focus on the shape of a trajectory, searching for characteristic points where the geometric structure of the trajectory changes substantially (Yoon and Shahabi 2008). Other approaches rather focus on derived movement parameters such as speed, heading, or sinuosity, and search for subtrajectories expressing uniform movement parameters (Buchin et al. 2010b, 2011b). Yet other work tries to understand the semantics of moves segmenting trajectories according to travel mode changes or moves between points of interest (Sester et al. 2012). Most of that work is rather methods-driven, its applications are hence mostly illustrative. In some CMA application areas, however, trajectory segmentation is closely related to applied research questions. For example in movement ecology, trajectory segmentation is sometimes referred to as classification such that the task is to semantically annotate segments of a trajectory with the most likely behavior of the observed animal. For instance, Shamoun-Baranes et al. (2012) use trajectory data in conjunction with additional sensory data from an accelerometer aiming at segmenting and labeling fixes according to a set of predefined behavior classes (here fly, forage, body care, stand, and sit) using supervised classification trees.

In the database community segmentation concepts have been suggested for structuring raw streams of position data into semantically meaningful units, aiming at supporting a meaningful interpretation of trajectories. In their conceptual view on trajectories Spaccapietra et al. (2008) base the semantic enrichment of raw movement data on an initial segmentation of trajectories into stops and moves. Subsequent semantic annotation could then label stops and moves with, for example, the type of activity (e.g., commute) or the type of the visited location (e.g., home vs. work). Baglioni and Fernandes de Macedo (2009) employ formal ontologies to enable such semantic enrichment, aiming at augmenting both the semantics of the trajectory data and also patterns mined from the trajectories. The authors argue that most pattern mining approaches produce patterns, one could say in a mechanistic way, that are then difficult to link to the actual movement behavior, i.e., the semantics (Baglioni et al. 2009, p. 272). By contrast, their semantic enrichment bases the interpretation of mined patterns in a domain ontology representing the geographical knowledge of the relevant application domain.

Similarity and clustering. There is ample related work on the similarity of trajectories. Some of these focus on the spatial or temporal characteristics of trajectories, or specifically aim at their spatio-temporal nature. From within computational geometry emerged a family of similarity measures based on the Fréchet distance between two curves (Buchin et al. 2010a). When a person on one curve walks a dog on the other curve, then the Fréchet distance is the shortest leash length that allows the person and the dog to walk their curves without either of them backtracking. Such basic measures can be extended to incorporate additional constraints, such as time correspondence, direction or speed. The same group of researchers also produced work on finding long and similar parts of trajectories (subtrajectories) based on a distance measure that is defined as the average Euclidean distance at corresponding times (Buchin et al. 2011a), and on context-aware similarity measures (Buchin et al. 2012).

Focusing on the temporal nature of trajectories work has been presented that adapts methods from time series analysis for assessing trajectory similarity. Similar to work on edit distance in Dodge et al. (**P14**. 2012), work has been produced on dynamic time warping (DTW, Vlachos et al. 2004) or longest common sub-sequence (LCSS, Vlachos et al. 2002a, b). Both DTW and LCSS allow for the detection of "elastic" matches, allowing stretching of sequences and giving more weight to similar portions of the sequences. Explicitly aiming at supporting knowledge discovery in moving object databases (MOD) Pelekis et al. (2012) propose several explicitly spatio-temporal trajectory similarity measures, based on primitive (space and time) as well as derived parameters of trajectories (speed, acceleration, and direction). Their querying framework bases on the distance measures *Locality In-between Polylines (LIP)*, which in essence computes the area of the shape formed in between two polylines.

Movement patterns. Following early work on flocking patterns (Laube et al. 2005) and subsequent work studying various flavors of the flocking problem and its efficient detection (Benkert et al. 2008), the concept of objects moving together in some form has seen several variants in related fields. In a database context for traffic applications, Jeung et al. (2008a, b) present a density-based notion of *convoys*. Here, instead of using an areal constraint (a disc of radius r), convoys consist of sets of entities that are density-connected, i.e., piece-wise linked given some distance threshold e. Huang et al. (2008) propose the notion of *herds* derived from density-based clustering, with herds allowing for non-spherical clusters and changing memberships. Finally for coherence mining in pedestrian data, Wachowicz et al. (2011) explicitly search for *moving flocks*, where in contrast to stationary flocks all members of the flock must *move* together for a certain period of time.

Similar to Merki and Laube (**P16**. 2012) Van de Weghe et al. (2006) and Gottfried (2011) also study interaction between moving objects. Van de Weghe and colleagues present the *Qualitative Trajectory Calculus* (QTC) for qualitative reasoning about the movement of pairs of moving point objects. Even though the authors present their QTC as a means for both representation and reasoning, I would argue that the major contribution lies in the reasoning part. Whereas other work on qualitative spatial reasoning has addressed static relations between entities (for example the *Region Connection Calculus*, RCC, in Randell et al. 1992), the QTC line of work specifically investigates changes between moving objects when there is no change in their topological relationship. Even though objects are and remain disconnected, one might want to reason if they move "towards each other", "away from each other", or are "stable with respect to each other". QTC represents such qualitative dynamic behavior of a pair of moving point objects using a small set of symbols. The authors illustrated their calculus for example for a predator-prey interaction in a 2D Euclidean space (Van de Weghe et al. 2006) or for movement along a road network (Van de Weghe et al. 2004; Bogaert et al. 2007). Gottfried (2011) argues in a very similar way that decomposing motion pattern into qualitative features and employing formal grammars has the advantage of being closer to human thinking and also suits often noisy and imprecise data. In his piece, however, he does not consider velocity or distance constraints, but investigates in contrast whether meaningful features can be derived from directional information alone.

Whereas much previous work in GIScience has studied movement patterns in isolation from their geographical embedding, more and more attention has recently been given to context-aware movement analysis. Orellana explores various aspects around pedestrian movement patterns as the result of the interactions between people and their environment (Orellana 2012). Orellana and Wachowicz (2011) and Orellana et al. (2012) propose the use of a local indicator of spatial association (LISA, Anselin 1995), a measure for assessing local spatial autocorrelation to detect spatial clusters of low speed vectors. Such *suspension* patterns explicitly do not search for stopping behavior in individuals' trajectories but rather for collective behavior, as in potential points of interest where many pedestrians stop (outlook, visitor center, picnic area). When ordered in *sequences*, frequently visited sets of such stopping clusters allows for the aggregation of visitor movement into flows.

Exploratory analysis and visual analytics. Introductory reviews for visual analytics of spatio-temporal data in general can be found in Andrienko et al. (2010) and specifically for movement data in Andrienko and Andrienko (2007). Given its explicit spatio-temporal character, movement data has served in exploratory analysis and visual analytics research as a signature case study. An extensive review of the wide range of literature on exploratory analysis and visual analytics of movement data would diverge too much from the path set out for this brief volume. Entry points for selected aspects of exploratory analysis and visual analytics of movement data can be found regarding generalization and aggregation (Andrienko and Andrienko 2011), density estimation and related concepts (Downs and Horner 2010, 2012), three-dimensional space-time cubes, combined with additional techniques for structuring the data (Rinzivillo et al. 2008; Demsar and Virrantaus 2010; Pelekis et al. 2012), interactive combinations of visualization and clustering (Schreck et al. 2009), as well as integrated visual analytics interfaces featuring linked and coordinated views of spatial, temporal, and socio-economic characteristics (Zhang et al. 2013).

Evaluation. Surprisingly little work has been produced on evaluating proposed move-ment mining methods. As mentioned in Sect. 3.3 it is difficult to get hold of suitable data featuring the semantic annotation needed for a thorough evaluation. However, there is work where an explicit focus was put making sure that the methods were sound and produced useful knowledge. One example for an assessing internal valid-ity using a test data set can be found in Orellana and Wachowicz (2011), where mined suspension patterns (a stopping behavior) are compared with reference or "ground truth" data. This reference data comes in the form of identified points of interest that the individuals are likely to visit and thereby express the suspension pattern: check-points in an outdoor game, as well as points of interest and crucial infrastructure in a recreational park. Cross-checking which of the expected points of interest are actually detected and which are missed allows an indication of error of omission and commission.

In problem-driven research domains the thorough evaluation of proposed methods is more common than in somewhat theoretical computer science or GIScience, also because in ecology a thorough evaluation is often a *condicio sine qua non* for publication. Given its roots in behavioral science, movement ecology occasionally produces movement with rich semantic annotation, painstakingly captured by human observers in the field. For instance in Shamoun-Baranes et al. (2012) the classification of oystercatcher behavior through supervised classification trees on location and sensor data was cross-checked with simultaneous visual observations. Similarly, Guilford et al. (2009) evaluate a machine learning based behavior classification (akin segmentation and labeling) through cross-validation of two simultaneously recorded sensor streams.

3.5 Concluding Remarks

The introduction of data mining concepts into GIScience with the goal of a better understanding of movement processes has clearly led to significant progress in structuring notoriously messy movement data. In an area dominated by a static view of the world inherited from cartography, the arrival of a flexible toolset allowing the search for patterns, trends, and similarities not only in space, but explicitly in space, time, and attributes was much needed and hence is to be warmly welcomed. Since movement data is inherently spatio-temporal, recording the location of an object at potentially thousands of time stamps can rapidly flood and "fill-up" maps, the GIScience signature analysis metaphor. Here, data mining's approach of conceptualizing and formalizing patterns and rules, akin search templates, that then can be searched for by efficient algorithms, offers analytical tools complementing the conventional GIScience tool box.

Many areas have contributed to establishing data mining and knowledge discovery in databases as a key toolsets of CMA. *GIScience* has contributed its theory of representing and abstracting both the moving entities as well as the spaces embedding movement. GIScience' theory on modeling spatio-temporal phenomena, entities and processes of the natural and built environment have made a significant contribution to the conceptualization of movement mining patterns and rules. A sometimes underestimated key contribution lies in GIScience' expertise in integrating multi-source and multi-scale data, preprocessing and transforming uncertain and noisy geodata to make it ready for the data mining algorithms. The *computational geometry* community has contributed in many ways to movement mining in CMA. Whereas early collaborations investigated movement patterns such as flock, convoy, leadership, the focus has recently shifted towards trajectory similarity, simplification and aggregation, as well as segmentation. The *database* community has made significant contributions to CMA with respect to storing, managing and querying movement data in specialized moving object databases (MOD) on the one hand, and by increasingly producing relevant research on data mining applications for movement data. The concept of semantic enrichment is another important concept towards structuring movement data streams for a better understanding of the underlying dynamic processes. Finally, *visual analytics* happily adopted the problem of movement analysis, allowing the efficient integration of all the above contributions in interactive analytics environments.

Hence, data mining helped CMA making significant progress in seeking structure in movement data, but semantic annotation of the found patterns remains difficult. Isolated analysis of the geometric footprint of movement is far from understanding movement behavior. Given the complexity of human and animal behavior, more and more researchers acknowledge that it may simply be too difficult to understand complex behavior just by studying its mere spatio-temporal footprint. It is thus little surprising that more and more work aims at capturing multi-sensor data, where location data is complemented by sensors that simultaneously record other attributes indicating the observed activity, such as acceleration, heart rate, or other physical properties of the monitored individual. Clearly, combining different sensor readings, where the location is only one variable, opens up exciting research avenues, advancing movement mining towards activity mining.

Despite initial work on ontological foundations of movement processes, there is little agreement in sight on a set of basic operations and patterns. For a start, the application domains interested in computational movement analysis seem to be so diverse, the phenomena they all study so variable that patterns or rules developed for one application simply have no relevance in another. Similarly, since the application problems are so diverse and the supply of new and interesting problems seems endless, the majority of movement mining methods remain custom-built prototypes tailored to one specific problem. A wider agreement on a basic set of movement analysis problems and their possible solutions—leading towards a theory—is rare so far.

Promoting data mining tools in GIScience for CMA clearly led to a much needed shift of perspective. On the downside, so far CMA as a developing area remains dominated by tool-driven researchers. In most cases it's not applications people requesting the involvement of methods people, but its methods people searching for interesting applications. This brings the danger of an analytical process that tries to reshape the real world to fit preexisting solutions. Database experts see the world of moving things in queries, computational geometry experts in data structures and algorithms, visual analytics experts in multi-views, parallel coordinate plots and space-time cubes. "If all you have is a hammer, everything looks like a nail". Community efforts such as the 2012 workshop on "Progress in movement analysis" held in Zurich, explicitly seeking research emerging collaborations between methods experts and domain specialists, acknowledge this danger.

References

Andersson, M., Gudmundsson, J., Laube, P., & Wolle, T. (2008). Reporting leaders and followers among trajectories of moving point objects. *GeoInformatica, 12*(4), 497–528.

Andrienko, G., Andrienko, N., Demsar, U., Dransch, D., Dykes, J., Fabrikant, S. I., et al. (2010). Space, time and visual analytics. *International Journal of Geographical Information Science, 24*(10), 1577–1600.

Andrienko, N., & Andrienko, G. (2007). Designing visual analytics methods for massive collections of movement data. *Cartographica, 42*(2), 117–138.

Andrienko, N., & Andrienko, G. (2011). Spatial generalization and aggregation of massive movement data. *IEEE Transactions on Visualization and Computer Graphics, 17*(2), 205–219.

Anselin, L. (1995). Local indicators of spatial association—LISA. *Geographical Analysis, 27*(2), 93–115.

Baglioni, M., & Fernandes de Macedo, J. A. (2009). Towards semantic interpretation of movement behavior advances in giscience. In M. Sester (Ed.), *Advances in GIScience* (pp. 271–288)., Lecture Notes in Geoinformation and Cartography Berlin: Springer.

Benkert, M., Gudmundsson, J., Hübner, F., & Wolle, T. (2008). Reporting flock patterns. *Computational Geometry, 41*(3), 111–125.

Bertin, J., Berg, W., and Scott, P. (1981). *Graphics and graphic information processing.* De Gruyter.

Bleisch, S., Duckham, M., Galton, A., Laube, P., & Lyon, J. (2014). Mining candidate causal relationships in movement patterns. *International Journal of Geographical Information Science, 28*(2), 363–382.

Bogaert, P., Van De Weghe, N., Cohn, A. G., Witlox, F., & De Maeyer, P. (2007). The qualitative trajectory calculus on networks. *Spatial cognition V reasoning, action, interaction* (Vol. 4387, pp. 20–38)., Lecture Notes in Computer Science, LNAI Berlin: Springer.

Both, A., Duckham, M., Laube, P., Wark, T., & Yeoman, J. (2013). Decentralized monitoring of moving objects in a transportation network augmented with checkpoints. *The Computer Journal, 56*(12), 1432–1449.

Buchin, K., Buchin, M., & Gudmundsson, J. (2010a). Constrained free space diagrams: A tool for trajectory analysis. *International Journal of Geographical Information Science, 24*(7), 1101–1125.

Buchin, K., Buchin, M., van Kreveld, M., & Luo, J. (2011a). Finding long and similar parts of trajectories. *Computational Geometry, 44*(9), 465–476.

Buchin, M., Dodge, S., Speckmann, B., et al. (2012). Context-aware similarity of trajectories. In N. Xiao, M. -P. Kwan, M. Goodchild, & S. Shekhar (Eds.), *Geographic information science*. Lecture Notes in Computer Science (Vol. 7478, pp. 43–56). Berlin: Springer.

Buchin, M., Driemel, A., van Kreveld, M., & Sacristan, V. (2010b). An algorithmic framework for segmenting trajectories based on spatio-temporal criteria. In 18th International Conference on Advances in Geographic Information Systems (ACM SIGSPATIAL GIS. (2010). *San Jose. California: ACM.*

Buchin, M., Driemel, A., van Kreveld, M., & Sacristan, V. (2011b). Segmenting trajectories: A framework and algorithms using spatiotemporal criteria. *JOSIS, 3*, 33–63.

Chakrabarti, S., Ester, M., Fayyad, U., Gehrke, J., Han, J., Morishita, S., & et al. (2006). *Data mining curriculum: A proposal.* Intensive Working Group of ACM SIGKDD Curriculum Committee: Technical report.

Demsar, U., & Virrantaus, K. (2010). Space-time density of trajectories: Exploring spatio-temporal patterns in movement data. *International Journal of Geographical Information Science, 24*(10), 1527–1542.

Dennis, T. E., Chen, W. C., Koefoed, I. M., Lacoursiere, C. J., Walker, M. M., Laube, P., et al. (2010). Performance characteristics of small global-positioning-system tracking collars for terrestrial animals. *Wildlife Biology in Practice, 6*(1), 14–31.

Dodge, S., Weibel, R., & Lautenschütz, A.-K. (2008). Towards a taxonomy of movement patterns. *Information Visualization, 7*(3–4), 240–252.

Dodge, S., Laube, P., & Weibel, R. (2012). Movement similarity assessment using symbolic representation of trajectories. *International Journal of Geographical Information Science, 26*(9), 1563–1588.

Downs, J. A., & Horner, M. W. (2010). In S. Fabrikant, T. Reichenbacher, M. Kreveld, & C. Schlieder (Eds.), *Geographic information science*. Lecture Notes in Computer Science (Vol. 6292, pp. 16–26). Berlin: Springer.

Downs, J. A., & Horner, M. W. (2012). Analysing infrequently sampled animal tracking data by incorporating generalized movement trajectories with kernel density estimation. *Computers, Environment and Urban Systems, 36*(4), 302–310.

Dumont, B., Boissy, A., Achard, C., Sibbald, A. M., & Erhard, H. W. (2005). Consistency of animal order in spontaneous group movements allows the measurement of leadership in a group of grazing heifers. *Applied Animal Behaviour Science, 95*(1–2), 55–66.

Fayyad, U., Piatetsky-Shapiro, G., & Smyth, P. (1996). From data mining to knowledge discovery in databases. *AI Magazine, 17*(3), 37–54.

Galton, A. (2005). Dynamic collectives and their collective dynamics. In A. Cohn & D. M. Mark (Eds.), *Spatial Information Theory, Proceedings*. Lecture Notes in Computer Science (Vol. 3693, pp. 300–315). Heidelberg: Springer.

Geng, L., & Hamilton, H. J. (2006). Interestingness measures for data mining: A survey. *ACM Computing Surveys, 38*(3), 9.

Gonzalez, M. C., Hidalgo, C. A., & Barabasi, A. L. (2008). Understanding individual human mobility patterns. *Nature, 453*(7196), 779–782.

Gottfried, B. (2011). Interpreting motion events of pairs of moving objects. *GeoInformatica, 15*(2), 247–271.

Guilford, T., Meade, J., Willis, J., Phillips, R., Boyle, D., Roberts, S., et al. (2009). Migration and stopover in a small pelagic seabird, the manx shearwater puffinus puffinus: Insights from machine learning. *Proceedings of the Royal Society B: Biological Sciences, 276*(1660), 1215–1223.

Han, J., & Kamber, M. (2006). *Data mining: Concepts and techniques*. Amsterdam: Morgan Kaufmann Publishers.

Hand, D. J., Manilla, H., & Smyth, P. (2001). *Principles of data mining*. Cambridge, MA: MIT Press.

Huang, Y., Chen, C. & Dong, P. (2008). Modeling herds and their evolvements from trajectory data. *Proceedings of Fifth International Conference on Geographic Information Science*.

Jeung, H., Shen, H. T., & Zhou, X. (2008a). Convoy queries in spatio-temporal databases. In *2008 IEEE 24th International Conference on Data Engineering* (pp. 1457–1459), Cancun, Mexico.

Jeung, H., Yiu, M. L., Zhou, X., Jensen, C. S., & Shen, H. T. (2008b). Discovery of convoys in trajectory databases. *Proceedings of the VLDB Endowment, 1*(1), 1068–1080.

Keim, D., Andrienko, G., Fekete, J.-D., Görg, C., Kohlhammer, J. & Melançon, G. (2008). Visual analytics: Definition, process, and challenges. In A. Kerren, J. Stasko, J.-D. Fekete, C. North (Eds.), *Information visualization*. Lecture Notes in Computer Science (Vol. 4950, pp. 154–175). Berlin: Springer.

Laube, P. (2009) Progress in movement pattern analysis. In B. Gottfried & H. Aghajan (Eds.), *Behaviour monitoring and interpretation, BMI, smart environments*. Ambient Intelligence and Smart Environments (Vol. 3, pp. 43–71). Amsterdam, NL: IOS Press.

Laube, P., Berg, M., Kreveld, M., et al. (2008a). Spatial support and spatial confidence for spatial association rules. In A. Ruas & C. Gold (Eds.), *Headway in spatial data handling*. Berlin: Springer.

Laube, P., Dennis, T., Walker, M., & Forer, P. (2007). Movement beyond the snapshot–dynamic analysis of geospatial lifelines. *Computers, Environment and Urban Systems, 31*(5), 481–501.

Laube, P., Duckham, M., & Palaniswami, M. (2011a). Deferred decentralized movement pattern mining for geosensor networks. *International Journal of Geographical Information Science, 25*(2), 273–292.

Laube, P., Duckham, M., & Wolle, T. (2008b). Decentralized movement pattern detection amongst mobile geosensor nodes. In T. J. Cova, K. Beard, M. F. Goodchild, & A. U. Frank (Eds.), *GIScience 2008*. Lecture Notes in Computer Science (Vol. 5266, pp. 199–216). Berlin: Springer.

Laube, P., Gottfried, B., Klippel, A., Billen, R., & van de Weghe, N. (2011b). Report on the first workshop on movement pattern analysis MPA10. *JOSIS*, *1*(2), 127–133.

Laube, P., & Purves, R. (2006). An approach to evaluating motion pattern detection techniques in spatio-temporal data. *Computers, Environment and Urban Systems*, *30*(3), 347–374.

Laube, P., & Purves, R. S. (2011). How fast is a cow? Cross-scale analysis of movement data. *Transactions in GIS*, *15*(3), 401–418.

Laube, P., van Kreveld, M., & Imfeld, S. (2005). Finding REMO–detecting relative motion patterns in geospatial lifelines. In P. F. Fisher (Ed.), *Developments in Spatial Data Handling, Proceedings of the 11th International Symposium on Spatial Data Handling* (pp. 201–214). Berlin, DE: Springer.

Merki, M., & Laube, P. (2012). Detecting reaction movement patterns in trajectory data. In J. Gensel, D. Josselin, & D. Vandenbroucke (Eds.), *AGILE'2012 International Conference on Geographic Information Science*. FR: Avignon.

Miller, H., & Han, J. (2009). *Geographic data mining and knowledge discovery*. Boca Raton: CRC Press.

Mohammad, Y., & Nishida, T. (2010). Mining causal relationships in multidimensional time series. In E. Szczerbicki & N. Nguyen (Eds.), *Smart information and knowledge management*. Studies in Computational Intelligence (Vol. 260, pp. 309–338). Berlin: Springer.

Nagy, M., Akos, Z., Biro, D., & Vicsek, T. (2010). Hierarchical group dynamics in pigeon flocks. *Nature*, *464*(7290), 890–893.

Orellana, D. (2012). Exploring Pedestrian Movement Patterns (PhD thesis, Wageningen University).

Orellana, D., Bregt, A. K., Ligtenberg, A., & Wachowicz, M. (2012). Exploring visitor movement patterns in natural recreational areas. *Tourism Management*, *33*(3), 672–682.

Orellana, D. & Renso, C. (2010). Developing an interactions ontology for characterising pedestrian movement behaviour. In *Movement-aware applications for sustainable mobility: Technologies and approaches* (pp. 62–86). IGI Global.

Orellana, D., & Wachowicz, M. (2011). Exploring patterns of movement suspension in pedestrian mobility. *Geographical Analysis*, *43*(3), 241–260.

Pelekis, N., Andrienko, G., Andrienko, N., Kopanakis, I., Marketos, G., & Theodoridis, Y. (2012). Visually exploring movement data via similarity-based analysis. *Journal of Intelligent Information Systems*, *38*(2), 343–391.

Peterson, R. O., Jacobs, A. K., Drummer, T. D., Mech, L. D., & Smith, D. W. (2002). Leadership behavior in relation to dominance and reproductive status in gray wolves. *Canis lupus. Canadian Journal of Zoology*, *80*(8), 1405–1412.

Randell, D. A., Cui, Z., & Cohn, A. G. (1992). A spatial logic based on regions and connection. *KR*, *92*, 165–176.

Richter, K.-F., Schmid, F., & Laube, P. (2012). Semantic trajectory compression: Representing urban movement in a nutshell. *JOSIS*, *4*, 3–30.

Rinzivillo, S., Pedreschi, D., Nanni, M., Giannotti, F., Andrienko, N., & Andrienko, G. (2008). Visually driven analysis of movement data by progressive clustering. *Information Visualization*, *7*(3–4), 225–239.

Rykiel, E. J. J. (1996). Testing ecological models: The meaning of validation. *Ecological Modelling*, *90*(3), 229–244.

Schreck, T., Bernard, J., von Landesberger, T., & Kohlhammer, J. (2009). Visual cluster analysis of trajectory data with interactive Kohonen maps. *Information Visualization*, *8*(1), 14–29.

Sester, M., Feuerhake, U., Kuntzsch, C., & Zhang, L. (2012). Revealing underlying structure and behaviour from movement data. *KI*, *26*(3), 223–231.

Shamoun-Baranes, J., Bom, R., van Loon, E. E., Ens, B. J., Oosterbeek, K., & Bouten, W. (2012a). From sensor data to animal behaviour: An oystercatcher example. *PLoS ONE*, *7*(5), e37997.

Shamoun-Baranes, J., van Loon, E. E., Purves, R. S., Speckmann, B., Weiskopf, D., & Camphuysen, C. J. (2012b). Analysis and visualization of animal movement. *Biology Letters*, *8*(1), 6–9.

Shapiro, L. G., & Stockman, G. C. (2001). *Computer vision*. New Jersey: Prentice-Hall.

Silberschatz, A., & Tuzhilin, A. (1996). What makes patterns interesting in knowledge discovery systems. *IEEE Transactions on Knowledge and Data Engineering, 8*(6), 970–974.

Spaccapietra, S., Parent, C., Damiani, M. L., de Macedo, J. A., Portoa, F., & Vangenot, C. (2008). A conceptual view on trajectories. *Data and Knowledge Engineering, 65*(1), 126–146.

Thomas, J. J., & Cook, K. A. (2006). A visual analytics agenda. *IEEE Computer Graphics and Applications, 26*(1), 10–13.

Tufte, E., & Graves-Morris, P. (1983). *The visual display of quantitative information* (Vol. 31). Cheshire, CT: Graphics Press.

Van de Weghe, N., Cohn, A. G., Bogaert, P., & De Maeyer, P. (2004). Representation of moving objects along a road network. In *Proceedings of the 12th International Conference on Geoinformatics*, Citeseer.

Vlachos, M., Gunopulos, D., & Das, G. (2004). Rotation invariant distance measures for trajectories. In *Proceedings of the 10th ACM SIGKDD International Conference on Knowledge Discovery and Data Mining* (pp. 707–712). Seattle, WA. ACM.

Vlachos, M., Gunopulos, D., & Kollios, G. (2002a). Robust similarity measures for mobile object trajectories. In *Proceedings of 13th International Workshop on Database and Expert Systems Applications* (pp. 721–728). IEEE Computer Society.

Vlachos, M., Kollios, G., & Gunopulos, D. (2002b). Discovering similar multidimensional trajectories. In *Proceedings of 18th International Converence on Data Engineering (ICDE'02)*.

Wachowicz, M., Ong, R., Renso, C., & Nanni, M. (2011). Finding moving flock patterns among pedestrians through collective coherence. *International Journal of Geographical Information Science, 25*(11), 1849–1864.

Van de Weghe, N., Cohn, A. G., De Tré, G., & De Maeyer, P. (2006). A qualitative trajectory calculus as a basis for representing moving objects in geographical information systems. *Control and Cybernetics, 35*(1), 97–119.

Wood, Z., & Galton, A. (2009a). Classifying collective motion. In B. Gottfried & H. Aghajan (Eds.), *Behaviour monitoring and interpretation–BMI–smart environments*. Ambient Intelligence and Smart Environments (Vol. 3, pp. 129–155). Amsterdam, NL: IOS Press.

Wood, Z., & Galton, A. (2009b). A taxonomy of collective phenomena. *Applied Ontology, 4*(3), 267–292.

Yoon, H. & Shahabi, C. (2008). Robust time-referenced segmentation of moving object trajectories. In *8th IEEE International Conference on Data Mining (ICDM '08)* (pp. 1121–1126).

Zhang, Q., Slingsby, A., Dykes, J., Wood, J., Kraak, M.-J., Blok, C. A., & Ahas, R. (2013). Visual analysis design to support research into movement and use of space in tallinn: A case study. *Information Visualization*. (In Press).

Chapter 4
Decentralized Movement Analysis

Current technological advancements have lead to an ever increasing integration of information and communication technologies. This chapter investigates what happens when movement analysis is no longer constrained to desktop information systems, but instead migrates into the highly dynamic network built by the mobile phones of commuters or the board computers of taxis in a fleet management system. The bidding war between Google and Facebook over the navigation app Waze may serve as an example for the increasing relevance of applications where mobile but networked devices are used for collecting and integrating spatial information (Dredge 2013). Waze integrates on the fly spatial knowledge about traffic flow and road conditions through a process of crowd-sourcing from drivers acting as mobile sensors.

This chapter combines Computational Movement Analysis (CMA) with decentralized spatial computing (DeSC), the theoretical underpinning of geosensor networks (GSN) and vehicular ad hoc networks (VANETs). As a special flavor of DeSC, the chapter explores the foundation of decentralized movement analysis. In doing so this chapter shifts focus as it addresses a rather technology-driven problem area. Instead of discussing conceptual models or analytical techniques irrespective of any underlying system architecture, it digs deeper and investigates the interplay between a specific system architecture and its superimposed data processing procedures.

The chapter not only investigates how movement analysis tasks similar to the ones discussed in the two previous chapters can still be performed in decentralized spatial information systems, but also investigates how the movement of tracked objects and tracking devices can be exploited in a wider sense for information processing in such systems. Specifically, this chapter investigates the following CMA tasks in a decentralized way: Movement pattern detection in Laube et al. (**P6**. 2008; **P12**. 2011), assessing the network load in a transportation network in Both et al. (**P19**. 2013), point clustering in Laube and Duckham (**P8**. 2009), and Laube et al. (**P11**. 2010), and privacy-safeguarding in location-based services in Laube et al. (**P11**. 2010).

Overarching research objectives. The research summarized in this chapter contributes to the following overarching research objectives of computational movement analysis.

© The Author(s) 2014

P. Laube, *Computational Movement Analysis*,
SpringerBriefs in Computer Science, DOI 10.1007/978-3-319-10268-9_4

- Characterize the limitations and opportunities of movement analysis in decentralized spatial information systems, and assess the role of CMA in such systems.
- Develop the theoretical underpinnings of information processing strategies that explicitly exploit movement of study objects and/or data capture devices in decentralized spatial information systems.

The chapter is structured as follows. First, Sect. 4.1 introduces the foundations of wireless sensor and geosensor networks building the basis for the subsequent discussion of decentralized movement analysis. Then, Sect. 4.2 discusses two settings where movement in decentralized spatial systems has been studied in the context of this book—static sensors monitoring passing mobile objects, and mobile agents autonomously monitoring their collective movement behavior. Finally, based on this discussion of two exemplary settings, Sect. 4.3 derives and isolates a set of more generic principles for decentralized movement analysis.

4.1 Foundations

Advances in distributed sensing and computing technologies offer new, reliable, and cost-effective means to collect fine-grained spatio-temporal information when monitoring natural and built environments—so-called wireless sensor networks. The definition *wireless sensor networks* (WSN) refers to wireless networks of untethered, battery powered miniaturized computers with the ability to sense, process, and communicate information in a collaborative way (Zhao and Guibas 2004). Example deployments include hazard management (Duckham et al. 2005), monitoring seismic activity (Werner-Allen et al. 2006) or traffic flow (Kellerer et al. 2001). When specifically monitoring phenomena in geographic space such systems are called geosensor networks (GSN, Nittel et al. 2004). Geosensor networks offer a powerful large-scale alternative to conventional small-scale remote sensing and ground surveys. Figure 4.1 illustrates the basic set-up of a geosensor network.

On top of the many technological challenges keeping the engineers busy, wireless sensor networks, and in the interest of this volume, geosensor networks also pose substantial challenges for processing the information generated in such networks. Whereas conventional geographic information processing is based on centralized computing models, where sophisticated and powerful databases collate and process information globally, no such omniscient computing capability exists in geosensor networks. By contrast, such systems require a new way of spatial computing, where spatial knowledge is generated from collaborating, but distributed computing nodes that have only partial knowledge (**P8**. Laube and Duckham 2009). Much conventional spatial computing nowadays is distributed computing (Worboys and Duckham 2004; Duckham 2012). That means that many information systems cooperate synchronously in order to complete a task. For example, for offering location-based services the mobile devices must communicate with a positioning system and spatial

Fig. 4.1 A geosensor
network consisting of sensor
nodes and edges of the
communication graph, given
a communication distance c.
The nodes measure
temperature. Nodes at the
boundary of the hot area
(*dashed line*) are shown in
black, other nodes in *white*

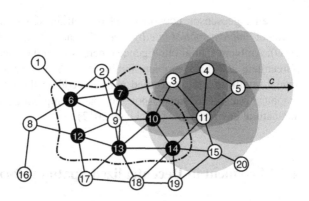

databases providing base data. However, in decentralized systems as a special case
of distributed systems, no single component knows the entire system state (Lynch
1996). In that sense, geosensor networks are decentralized systems as nodes spread
in space, sense their local environment, but must collaborate in order to generate
the complete picture of a monitored geographic phenomenon. To this end, Duckham
(2012, p. 16) defines *decentralized spatial computing* (DeSC) as "the study of the
decentralized algorithms, data structures, and technologies for computing with spatial information."

The reasons explaining the need for decentralized spatial computing are manifold
(**P8**. Laube and Duckham 2009; Duckham 2012).

- *Information overload.* Geosensor networks may consist of thousands of sensors
  sampling geospace at very fine temporal granularities. Here, decentralized spatial
  computing helps managing the potentially very large and highly autocorrelated
  data volumes generated in geosensor networks by filtering and processing the data
  in the network.
- *Scalability.* As networks scale from hundreds to thousands and potentially millions
  of nodes, centralized control of the system becomes impossible. Since decentralized systems are controlled through interactions between individuals, adding more
  nodes to the systems is simple, as the controlling rules can remain the same.
- *Sensor/actuator networks.* Information generated by a sensor node may often be
  required by other nodes close by (e.g., controlling irrigation through a humidity
  sensor network). What is sensed locally, matters locally. Removing information
  from the network, processing it centrally, and then returning it into the network
  would present an inefficient drain of network resources.
- *Latency.* Decentralized information processing requires communication and collaboration, which in turn takes time. Decentralized spatial computing can decrease
  the latency of the system.
- *Privacy.* Whereas centralized databases can represent a potential security
  breach, decentralized protocols can ensure that no single component can accumulate knowledge about any individual, hence privacy can be protected (**P11**.
  Laube et al. 2010).

Finally, geosensor networks and decentralized spatial computing lead to the vision of *ambient spatial intelligence* (**P7**. Laube et al. 2009; Duckham and Bennett 2009). Ambient spatial intelligence emerges from the idea of ubiquitous computing (Greenfield 2006) and ambient intelligence (Augusto and Shapiro 2007). Ambient spatial intelligence is concerned with embedding the intelligence to monitor geographical phenomena and respond to spatio-temporal queries directly into our built and natural environments (Duckham 2012).

4.2 Movement in Decentralized Spatial Information Systems

Movement is found in various forms in geosensor networks. Geosensor networks can monitor the movement of entities or mobility and dynamic processes in fields. Alternatively, the sensing nodes themselves can move and monitor a geographic phenomenon whilst moving. Mobile nodes could be passively suspended in a dynamic medium (i.e., nodes floating in a water current), carried by a mobile agent (i.e., a smart phone), or even have their own locomotion capabilities (i.e., autonomous agents in robotics). Finally, information tokens often move through the network, for instance when routing information from a source node to a sink. Hence, mobility can be the primary study subject, a property of the decentralized system, or a property of information that is handed around through the system amongst communicating and collaborating nodes. On all these levels, mobility presents at the same time challenges and opportunities. This section investigates what these challenges and opportunities are.[1]

Section 2.2 introduced three dimensions discriminating conceptual movement spaces and the therein embedded movement. Very similar consideration are relevant when investigating mobility in geosensor networks, except that with the *mobility mode* an additional fourth dimension should be considered here. The following list results from integrating the list in Sect. 2.2 with characteristics of mobile objects listed in Duckham (2012, Sect. 6.1.2, p. 173).

- *Constraints to movement.* Do the moving objects roam freely in Euclidean space or are they constrained by a transportation network?
- *Continuous versus discrete movement space.* How is the movement tracked, is it a quasi-continuous stream of position fixes captured with coordinates (GPS positions) or sequences of visited partitions of space (mobile phone antenna cells, network edges)?

---

[1] This book is solely about the analysis of movement data. Even though the distinction between data capture and data analysis gets occasionally a bit blurred in this chapter, there are important aspects of wireless sensor networks involving movement that are not covered in this chapter. For example, target tracking, that is in essence the capturing of raw positional data of moving objects, is not covered. Also information routing, another wireless sensor network classic, contributes to setting up and maintaining the network infrastructure, but is not considered analysis. Readers interested in such issues are referred to the introductory text on wireless sensor networks in Zhao and Guibas (2004).

Table 4.1 Mobility mode

		Sensor nodes	
		Static	Mobile
Phenomena	Static	I. Geosensor networks	II. Cluster mining with information grazing
	Mobile	III. Monitor flows in movement network, Sect. 4.2.1	IV. Decentralized flock detection, Sect. 4.2.2

What is mobile? The sensor nodes, the studied entities, or both?

- *Lagrangian versus Eulerian perspective.* Is the movement monitored as positions over time (trajectories) or as times when the object passes fixed checkpoints or cordons?
- *Mobility mode.* What is mobile? Are static nodes tracking mobile objects, or are mobile nodes monitoring a static environment, or are mobile nodes monitoring a dynamic environment?

Whereas the three dimensions set out in Sect. 2.2 only focus on the studied movement process and its embedding in a conceptual space model, in decentralized spatial information systems also the data capture system (i.e. a geosensor network or a VANET) can be subject to mobility. Since both the study object and the monitoring system can be on the move, there are four possible mobility modes to be considered for dynamic decentralized spatial information systems (Table 4.1). Except the static-static combination of conventional geosensor networks, all possible combinations were addressed in this book, with a focus on modes III and IV. Obviously, the more dynamic the system, the more difficult become the tasks. Hence, mode IV is expected to be more challenging than modes II and III (Duckham 2012).

Following the chapter on "Monitoring Spatial Change over Time" in Duckham (2012), two critical distinctions with respect to the information generated in a dynamic decentralized spatial information system shall be discussed in more detail here. The first one refers to the temporal nature of the generated information: Some systems record histories, others record chronicles (Galton 2004; Duckham 2012). *Histories* provide a spatio-temporal record of the states of monitored endurants (e.g. moving objects) through time.[2] By contrast, *chronicles* provide a record of the occurrences (perdurants) that happened through time (the occurrence of an object changing from transit edge e_i to e_j). This is important because most wireless sensor systems and hence also dynamic decentralized spatial information systems monitor histories (snapshots) (Duckham 2012). Therefore in systems requiring chronicles, occurrences will need to be inferred from states (Duckham 2012). It is this semantic enrichment that offers an opportunity for geographically informed algorithms in decentralized movement analysis discussed in this chapter.

---

[2] Recall the SNAP and SPAN ontologies (Grenon and Smith 2004). *Endurants* or *continuants* are things that endure through time, e.g. a moving object, this printed book (SNAP ontology). *Perdurants* or *occurrents* by contrast are things that occur in time, e.g. the reader reading this book (SPAN ontology).

The second distinction addresses the question whether the system only generates information or whether the system allows also for querying this information. This distinction also addresses the question where in the system the information is stored. Often, decentralized systems will have separate algorithms for first generating information or collating information in specific nodes in the network. The querying of this information will be delegated to separate algorithms (Duckham 2012). For example, in Both et al. (**P19**. 2013) most algorithms are concerned with the collation of information about the flow in the network in the cordons. Specific algorithms presented in the final section of the paper then investigate how this system would handle queries put to the monitoring system.

In the following two decentralized movement analysis tasks illustrate two opposite system architectures and related problems in the light of the general issues raised in this section. The examples include the decentralized monitoring of network flows in a cordon-structured network (mode III, Sect. 4.2.1) and the decentralized detection of flock movement patterns (mode IV, Sect. 4.2.2).

4.2.1 Static Nodes Monitor Mobile Objects

Unconstrained movement is rare. People move, for example, mostly in constrained transportation networks. Hence, infrastructure enabling, managing and monitoring network-bound movement becomes an important source for large movement data volumes. Examples range from cellular networks for mobile phones to electronic ticketing for public transport (e.g. London's Oyster card or Melbourne's myki system) and road tolling systems. In the context of this chapter such systems are interesting since some capture information in a decentralized way, where cars are observed when passing GSM towers or traffic cordons or commuters swiping card readers when hopping on and off trains or buses. Collating all that information capturing the whereabouts of agents in such systems in centralized databases may at best be impractical, in some cases simply impossible. Hence such systems lend themselves to decentralized spatial computing and decentralized movement analysis.

Both et al. (**P19**. 2013) present a family of algorithms for the decentralized monitoring of moving objects in such a transportation network augmented with checkpoints.[3] The approach considers mobile objects that move and are tracked on a *transportation network*, modeled as a graph where transportation edges connect intersections. The moving objects are tracked whenever they pass a cordon (or checkpoint) that are typically but not necessarily positioned at intersections. Figure 4.2 depicts a generalized ring-shaped network with four cordons indeed positioned at the intersections. The movement in this architecture is *constrained* to a

---

[3] In Both et al. (**P19**. 2013) "fish" is used as a shorthand for moving objects because the work was initiated in response to a set of problems coming out of a river health monitoring system deployed in the Murray River, Australia, tracking real fish with RF transmitters and riverside cordons (Koehn et al. 2008).

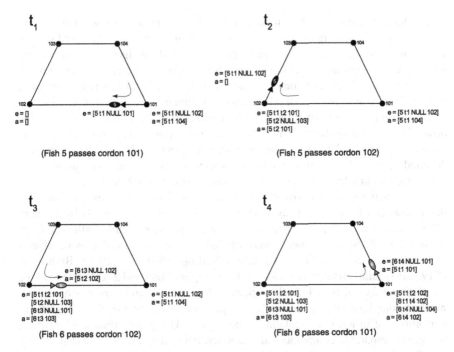

Fig. 4.2 Mobile objects ("fish") moving on a transportation network and passing cordons (Republished from Both, A., Duckham, M., Laube, P., Wark, T., & Yeoman, J. (2013). Decentralized monitoring of moving objects in a transportation network augmented with checkpoints. *The Computer Journal, 56*(12), 1432–1449, DOI:10.1093/comjnl/bxs117, by permission of Oxford University Press.)

discrete cordon-structured network. Furthermore this systems adheres to the *Eulerian* perspective of movement as the system monitors times when the objects pass the fixed cordons. The raw information produced by this tracking system consists of *histories* of the states of the moving endurants (i.e. object o is located on edge e_i). Just as outlined above, deriving chronicles of transition occurrences requires information processing.

On top of the transportation network, the formal model furthermore requires a communication network and a connectivity network. The model assumes one-hop communications between two types of sensor nodes: the set of moving nodes (fish), and the set of immobile cordons at known locations. The *communication network* is then represented as a time-varying undirected graph comprising the one-hop communication links between nearby nodes. As will be shown below, depending on the application and technologies used these links will vary. Finally, the *connectivity network* represents the relative network locations of cordons in terms of transportation network connectivity between cordon locations. Cordons can sense the movement of passing fish. For example in Fig. 4.2, node 102 senses at t_2 that fish 5 passes, coming from node 101 (where it left at t_1) and is now heading towards node 103. Finally, fish may also be able to sense other fish when they pass each other (not depicted in Fig. 4.2).

The decentralized algorithms for monitoring movement involves two stages—first maintaining and exchanging records about movement events at cordons or fish, and second querying these records. The paper presents a family of algorithms increasingly relaxing constraints about the communication network. Whereas first all cordons are connected, this assumption is relaxed as fish act as data mules to physically move information in absence of other communication links. This is shown in Fig. 4.2 where fish 6 carries information about fish 5 back to node 101 from t_3 to t_4. The principle of exploiting the movement of objects to move around information is termed *mobility diffusion* and will further be investigated in Sect. 4.3.2.2. The final stage (not depicted in Fig. 4.2) investigates what happens when fish passing one another hand over information tokens. All three algorithms are summarized in Table 4.2.

The algorithms were evaluated with respect to *scalability* and *latency*, that is the delay between an event occurring and the event being detected correctly by the algorithm. Latency is another decentralized spatial computing principle further explored in Sect. 4.3.3.3. All experiments were carried out with simulated moving objects in the agent-based modeling environment NetLogo (Wilensky 1999). Both et al. (**P19**. 2013) summarize the results as follows. Algorithms 1 and 2 scale linearly with the total number of fish. Equally expected, Algorithm 3 lends towards a communication complexity that scales with the square of the number of fish. Algorithm 1 performs best in terms of scalability and latency. Using data mules, Algorithm 2 achieves comparable computational efficiency, but at the cost of increased latency. Finally, Algorithm 3 can further reduce the latency of Algorithm 2, but for the cost of increased computational complexity (Table 4.3).

The structures presented in Both et al. (**P19**. 2013) also support algorithms for queries of variable complexity. The complexity of the queries depends on the required degree of coordination between the cordons. Whereas queries about the identity of individual fish on an edge or total edge load at a given time can be handled by individual cordons, queries about composite fish paths and collective movement of fish (akin to flocking patterns) require cordon collaboration. Collaboration will be further explored in Sect. 4.3.3.1.

Table 4.2 Three different decentralized algorithms monitoring movement in a cordon structured network in Both et al. (**P19**. 2013)

	Algorithm description	DeSC peculiarities
Algorithm 1	Basic algorithm where all cordons are directly connected in the communication network to cordon neighbors in the transportation graph	Best performance in terms of scalability and latency
Algorithm 2	Mule algorithm, where fish transport exit records back to cordons, Fig. 4.2	Mules compensate for disconnected cordons, but at cost of increased latency
Algorithm 3	Extended mule algorithm, where fish transport and exchange exit records	Handshakes decrease latency at cost of increased overall computational complexity

Table 4.3 Decentralized algorithms monitoring flock movement patterns

	Algorithm description	DeSC peculiarities
FLAGS	Flock patterns are inferred from maturing information tokens that survive constant exchange and validation, Fig. 4.3	Mobility diffusion, here handing around information tokens, allows for decentralized
DDIG	Latency allows individuals to accumulate spatial information, constant rearrangement offers opportunities for information exchange and enrichment, Figs. 4.4 and 4.5	Latency reduces detection error, but this effect wears off with long latencies

4.2.2 Mobile Agents Monitor Their Collective Movement

The second system architecture included in this chapter involves roaming sensor nodes that aim to infer movement patterns in a collaborative manner without the need for a centralized database. Both, Laube et al. (**P6**. 2008; **P12**. 2011) investigate if and how movement pattern mining can be performed in a decentralized spatial information system. Can mobile agents that sense their own location and the presence of neighbors infer if they build the movement pattern flock? Again, an (n, k, p)-flock is a form of collective movement and defined as set of n moving entities that stay within a disk of radius p for k time steps.

All algorithms summarized in this section are based on the assumption that the nodes move in an *unconstrained* space. Whereas Laube et al. (**P12**. 2011) relies on nodes that can record their own position as coordinates, Laube et al. (**P6**. 2008) only requires sets of detected neighbors. In both studies, positions and neighbors are sensed in a quasi continuous manner, be it with a GPS receiver, any other localization technology. Irrespective of the localization and sensing capabilities, both studies adhere to the *Lagrangian* perspective. The mobility mode is rather challenging since both the sensing nodes and the investigated phenomena (here the flock patterns) potentially move. The constant rearrangement of the network topology and with it all connectivity links requires alternative strategies to popular strategies in decentralized spatial computing such as trees or any other precomputed and maintained data structure.

First, Laube et al. (**P6**. 2008) illustrates how collaborating nodes can extend their knowledge beyond their limited individual spatial perception area. In the FLAGS algorithm (flocking amongst geosensors), information tokens capturing a list of candidate nodes forming a flock are exchanged between roaming sensor nodes (Fig. 4.3). At each time step, nodes in the flock validate if the flock persists, update their token accordingly, and pass it on to their immediate neighbors. Since invalid tokens are removed, tokens that persist for k time steps "flag" the presence of a flock pattern. The constant process of exchanging and validating information allows individual sensor nodes to learn from their neighbors about processes beyond their own limited perception range. The algorithm separates knowledge from sensor nodes, knowledge becomes mobile. Hence, FLAGS presents both a form of *mobility diffusion* (Sect. 4.3.2.2) and *separation* (Sect. 4.3.3.4).

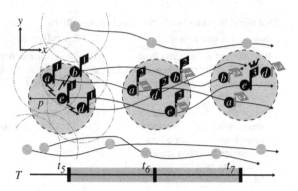

Fig. 4.3 Maturing knowledge tokens detect a ($n = 4, k = 3, p$)-flock locally. At t_5 sensor node e counts the required 4 neighbors, creates a token ($\{a, b, d, e\}, 1$) that is transmitted to all its neighbors within communication range c (little numbered flags). At t_6, after having moved and potentially rearranged, all sensor nodes check their tokens. This time only sensor node d counts enough neighbors and ages his token to ($\{a, b, d, e\}, 2$). All other sensor nodes drop their token. Sensor node d, however, forwards its aged token to all its neighbors. Finally, at t_7, again e counts enough neighbors, its token ($\{a, b, d, e\}, 3$) reaches the mature age $k = 3$, and flags a "found pattern"-message (*crown*) (With kind permission from Springer Science+Business Media: Cova, T. J., Beard, K., Good-child, M.F., & Frank, A.U. (Eds.). (2008). Geographic Information Science. *Lecture Notes in Computer Science, 5266*, ISBN 978-3-540-87472-0; Laube, P., Duckham, M., & Wolle, T. Decentralised movement pattern detection amongst mobil geosensor nodes, 199–216, Fig. 2.)

A similar form of mobility diffusion builds the core of the DDIG algorithm (deferred decentralized information grazing) presented in Laube et al. (**P12**. 2011). Here, roaming nodes exchange local histories of their recent positions whenever they get into each other's communication range (Fig. 4.4). Hence, through their mobility and thereby constantly changing communication partners, they build up local spatial knowledge that reaches beyond their individual spatial perception range (labeled "information grazing"). In a subsequent step, which is deferred by the time the information grazing requires, the local memory serves as a local knowledge base for the actual detection of the flock patterns (Fig. 4.5). To this end, any flocking algorithm could be used, in Laube et al. (**P12**. 2011) a simple heuristic was used based on an outlier elimination that suited the memory data structure.

Rather than investigating computational efficiency, both articles base their evaluation on an error analysis (see Sect. 3.3). The error analysis needs a data set where the number of detectable patterns is known. To this end simulated and real observation data was used. Then the decentralized algorithms were confronted with a set of experiments that systematically make their task harder, for instance, by reducing the communication radius (FLAGS and DDIG) or shortening the delay (DDIG). The error analysis then measured *error of omission* ("missed patterns") and *error of commission* ("false positives").

Both articles made clear that decentralized movement pattern mining is indeed possible, when imperfect, approximate, or delayed solutions are acceptable. They also showed that in the tested scenarios, node mobility was useful for succeeding

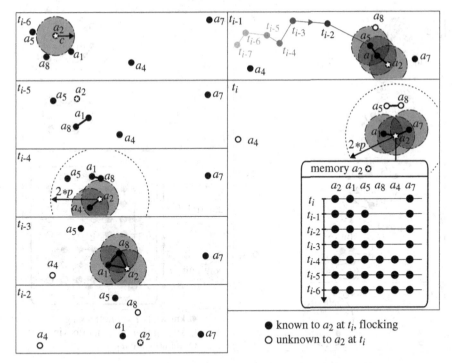

Fig. 4.4 DDIG builds a memory of the current movement history of nearby nodes. Example sensor node a_2 with communication range c roams from t_{i-6} to t_i and encounters nearby nodes. Whenever a_2 is connected to nearby nodes within $2*p$ ($2*p$-disc is only illustrated for t_i), it pulls their position history. a_2 received histories from a_4 at t_{i-4}, from a_8 at t_{t-3}, from a_5 at t_{i-1} (entire history illustrated for t_{i-1}), and finally from a_1 and a_7 at t_i. Memory a_2 reflects the corresponding build-up of local knowledge. Note, a_2 at t_i remembers previous positions of nodes it is currently not connected to (a_4, a_5, and a_8) (Republished from Laube, P., Duckham, M., & Palaniswami, M. (2011). Deferred Decentralized Movement Pattern Mining for Geosensor Networks. *International Journal of Geographical Information Science, 25*(2), 273–292, Taylor & Francis, DOI:10.1080/13658810903296630.)

in the given tasks. Imperfection came in the form of error of omission for small communication ranges. Imperfection also came in the form of latency (deferred processing) allowing the nodes to collect information and collaborate for its processing. DDIG showed that both in simulated and real data scenarios, detection error could be reduced by longer latency times. Latency is discussed in Sect. 4.3.3.3.

4.3 Decentralized Movement Analysis Principles

This section revisits the research included in this book and enumerates a set of information processing principles that illustrate the opportunities for decentralized spatial computing arising from mobility in geosensor networks. The principles listed here take advantage of properties of the studied phenomena or the models used for their

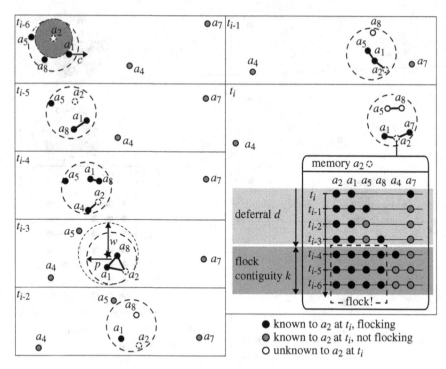

Fig. 4.5 The deferred data mining in DDIG. A heuristic is used to find flocks in the local memory a_2. A flock is detected for $[t_{i-6}, \ldots, t_{i-4}]$. Note that, given the deferred processing and the information stored in memory, a_2 finds flocks even with neighbors it is not connected anymore at t_i (Republished from Laube, P., Duckham, M., & Palaniswami, M. (2011). Deferred Decentralized Movement Pattern Mining for Geosensor Networks. *International Journal of Geographical Information Science, 25*(2), 273–292, Taylor & Francis, DOI:10.1080/13658810903296630.)

representation. Any preexisting knowledge can be exploited that contributes to solving a decentralized spatial computing task with its rather demanding preliminaries. Such knowledge can involve constraints to the movement (for example, nodes must move on the edges of a transportation network) or properties of the monitored phenomenon (for example, spatio-temporal autocorrelation of the movement parameters of flocking animals).

The following discussion illustrates that in decentralized spatial computing the boundaries between the previously separated tasks *data capture, communication* and *computation* become increasingly blurred (**P8**. Laube and Duckham 2009; Kargupta and Chan 2000). After a short recapitulation of the threats facing decentralized movement analysis, four opportunities arising mobility in geosensor networks are listed and exemplified in the context of decentralized movement analysis. The section concludes with a short discussion of the implications of these opportunities for decentralized movement analysis with respect to decentralization issues not specific to movement.

4.3.1 Challenges

Duckham (2012) suggests *neighborhood-based* or *location-based* decentralized spatial computing algorithms. Whereas neighborhood-based algorithms have only access to minimal spatial information in the form of the identities of neighbors, location-based algorithms exploit more complex spatial information such as direction, distance, and location of nodes. Movement challenges the underlying assumptions for both types of decentralized algorithms.

First, basic preliminaries about the spatial distribution of nodes and assumptions about the resulting communication graphs do no longer hold. When nodes move around, their spatial arrangement can be very heterogeneous, resulting in clusters of dense deployment separated by large empty gaps. This holds specifically when mobile nodes interact, hence meet, or are constrained to some form of transportation infrastructure. Establishing a communication network is difficult when the nodes show a heterogeneous and potentially unfavorable distribution.

Second, roaming nodes result in constantly changing topological node constellations regarding neighborhood, and changes in distances and direction. Apart from the danger of temporarily broken communication links, building up and maintaining data structures with enriched information about neighborhood, distances, or directions is at best difficult and often impossible. Nodes that are now close by neighbors move on and can be far away after a short while. Also, many DeSC algorithms function based on Tobler's first law of geography, assuming that the sensor readings of neighboring nodes tend to be similar. This assumption is also challenged, since both the monitored phenomena and the sensing nodes may move about. In essence, spatial structure and contiguity are no more static but potentially change constantly. Interacting, exchanging, and enriching information is even more difficult than in conventional geosensor networks.

Finally, when both the sensing system and the monitored phenomena can be mobile, it is difficult to know if changing sensor readings are due to an actual change in the monitored natural or built environment or simply a result of the sensing node getting a different perspective of an actually unchanged phenomenon.

4.3.2 Specific Decentralized Movement Analysis Principles

4.3.2.1 Mobility Compensation

Mobility compensation is a strategy for enlarging the reach of a sensor node. Nodes can be limited with respect to the perception area of their sensor (e.g., measuring the temperature around a node) or with respect to their communication range (maximal distance to maintain a communication link with a neighbor). Both, perception and communication range are often modeled and approximated with a disk around the sensor of a given radius. In static geosensor networks, individual sensor nodes might

Table 4.4 Decentralized CMA tasks addressed in this book and the respective decentralized movement analysis principles used

Task	Reference	Principle	Mode	Space
Movement pattern detection	Laube et al. (**P6**. 2008)	m.diffusion,	Moving nodes	Unconstrained
Movement pattern detection	Laube et al. (**P12**. 2011)	m.compensation, m.diffusion	Moving nodes	Unconstrained
Movement pattern detection	Both et al. (**P19**. 2013)	m.compensation, m.diffusion	Cordons	Network
Network load	Both et al. (**P19**. 2013)	m.diffusion	Cordons	Network
Clustering of point clusters	Laube and Duckham (**P8**. 2009)	m.compensation	Moving nodes	Unconstrained
Privacy-safe-guarding in LBS	Laube et al. (**P11**. 2010)	m.compensation, m.privacy, m.resilience	Moving nodes	Network

increase their spatial reach for sensing through collaboration with neighboring nodes. When the nodes are mobile, however, they can compensate their limited spatial reach simply through roaming and accumulating and integrating the gathered information.[4] Laube et al. (**P12**. 2011) called this process "information grazing". Even though in this very example the gathered information was purely the position history of the moving nodes, other applications could collect information about other spatial variables, such as, for example, humidity or temperature. Node mobility, allowing nodes to move about in space and time, allows for collecting and accumulating more information than could be accessed through a spatially limited perception range alone. It is important to note the a node can autonomously pursue mobility compensation, whilst most following strategies require collaborating nodes (Table 4.4).

4.3.2.2 Mobility Diffusion

Mobility diffusion is an important communication strategy in dynamic decentralized spatial information systems. *Mobility diffusion* refers to the strategy that moving nodes are allowed to physically move information around in the absence of connectivity (Grossglauser and Tse 2002; Grossglauser and Vetterli 2006; Duckham 2012). Thus, for instance, in Laube et al. (**P6**. 2008) roaming nodes hand over information tokens allowing for a repeated evaluation of the group composition leading towards the discovery of flock patterns. Knowledge is here "separated" from nodes, such that knowledge is no longer limited by the limitations of the nodes. Similarly, Laube et al. (**P12**. 2011) explore a combination of mobility compensation and mobility diffusion.

---

[4] Think of Pac-Man moving through his maze and eating away the pellets.

First, compensation leads to the build up of individual position histories, second these histories are handed around through handshakes made possible through diffusion.

Whereas mobility diffusion has the advantage of potentially bridging longer gaps between nodes with limited communication capabilities, this advantage comes at the cost of latency and the lack of delivery guarantee. Both, Grossglauser and Vetterli (2006) and Laube et al. (**P6**. 2008) emphasize that the movement regime of the moving nodes has a profound influence on the reliability of the mobility diffusion. The number of encounters clearly depends on the density of the original node deployment and their behavior. Only if nodes are expected to meet at all can this strategy offer an advantage. Finally, Laube et al. (**P11**. 2010) explicitly investigated the influence of different movement regimes in an application aiming at safeguarding privacy in an LBS scenario. Here, the balance between level of privacy and quality of service changed when replacing random walk with goal directed movement.

4.3.2.3 Mobility Privacy

In pre-internet and pre-database times privacy was naturally safeguarded through the fragmented nature of personal information sources (Rule et al. 1980). In essence, this early statement about privacy identifies the absence of a centralized database as the best strategy for safeguarding privacy. Since decentralized systems per definition do not require a centralized omniscient database, they naturally lend themselves to offering privacy protecting services. For example, a buddy tracker application could inform a user only through local communication when a friend enters his neighborhood, explicitly hiding both the location of the user and the friend from any centralized database.[5]

Laube et al. (**P11**. 2010) explore possible benefits that decentralization offers for safeguarding the privacy of LBS users. The paper studies this trade-off in a set of consecutive experiments simulating mobile and communicating agents roaming through a real urban transportation network. The experiments vary the used communication strategies (one-hop vs. multi-hop, push vs. pull services) and vary communication ranges. The results clearly indicate that mobility privacy is a valid strategy for LBS. For example, for the LBS query "Where is the nearest point of interest?", most relevant information naturally can be expected to reside in close proximity to the query agent. Since increasing hops and communication range leads to diminishing returns of increased quality of service, there is no need to decrease privacy through increased hops and communication radius.

The paper furthermore discusses the notion of *trajectory privacy* compared to conventional location privacy (**P11**. Laube et al. 2010). The idea here is, that for users of ICT services it might be perfectly acceptable to disclose the odd static location fix,

---

[5] Clearly, in most current ICT applications the system provider maintains a detailed log of the whereabouts and activities of its customers, but from a conceptual point of view underlining the argument of the *mobility privacy* opportunity such an omniscient system provider database is not a necessity.

since these are arguably erratic and quickly out of date. By contrast, the collection of entire trajectories allowing for detecting spatially explicit and potentially sensitive movement patterns poses a real privacy threat. Again, mobility privacy offers a solution as sensitive knowledge is "smeared" around in space-time such that no single system component can accumulate detailed knowledge about any individual.

4.3.2.4 Mobility Resilience

Network *resilience* is the ability of a network to defend against and maintain an acceptable level of service in the presence of malicious attacks as well as software and hardware failures (Smith et al. 2011). Clearly, resilience is also a key challenge for a sensor network consisting of cheap nodes in an ad-hoc network with potentially little influence on the spatial arrangement of the nodes and exposed to inhospitable environments. As outlined in the Sect. 4.3.1, the movement of nodes brings the danger of disconnected communication links.

Even though resilience has not been explicitly studied in the research summarized in this book, there are a few potential links to resilience. One could argue that mobility could help reconfigure a disconnected network just through rearrangement of nodes and hence overcome network failure. Even though not really investigated experimentally, the discussion in Laube et al. (**P11**. 2010) refers to scenarios with high turn-overs of POIs. It is then argued that mobility of nodes allows for a quick response to a constantly changing topology, underlining the intimate relation of proximity and decentralization. When new WiFi-hot spots show up in the neighborhood of a roaming agent, mobility helps discovering that new hot spot.

4.3.3 Revisiting General DeSC Principles

4.3.3.1 Collaboration

Collaboration between nodes is not particular to decentralized movement mining but essential for DeSC in general, as per definition no single node knows the entire system state (Lynch 1996) and is hence rarely able to solve a task alone. An obvious form of collaboration is information routing, where information packages hop from node to node. For instance, in Laube and Duckham (**P8**. 2009) nodes use multi-hop communication to ask their neighbors and their neighbors' neighbors if they are also sensing hot temperatures, hence they detect a cluster in a team effort.

Collaboration is found in many forms in decentralized movement analysis. The roaming nodes in Laube et al. (**P6**. 2008) hand around information tokens and validate thus in a collaborative way the persistence of a movement pattern in space-time. Here, collaboration aims at *mobility compensation*. Similarly, the nodes in Laube et al. (**P12**. 2011) addressing the same task of detecting flock movement patterns exchange information they have collected and in a collaborative effort complete each other's local neighbor memory required for the actual deferred data mining task. In both cases, mobility is essential as it leads to new node-node combinations and hence a recombination of opportunities for exchanging information. Finally, the "fish" in Both et al. (**P19**. 2013) collaborate since nurturing the memories of the cordons requires fish acting as data mules delivering information packages from other near-by cordons. All of the above examples also illustrate the collaborative nature of *mobility diffusion* and *mobility resilience*.

4.3.3.2 Local Relevance of Locally Sensed Information

In many geosensor systems information that is sensed by spatially distributed nodes will be most relevant to other nodes in their immediate spatial vicinity. This is especially the case in *sensor/actuator networks* where the locally sensed and processed information is also required locally by the network, for instance for a local activation of irrigation nozzles based on local sensor readings (**P8**. Laube and Duckham 2009; Duckham 2012). This principle—what is sensed locally is relevant locally—shows up repeatedly in the research summarized in this chapter. In decentralized flock detection algorithms nodes limit communication to other nodes in their vicinity and but not the other nodes in the network (e.g., **P12**. Laube et al. 2011).

In the LBS application in Laube et al. (**P11**. 2010) this local relevance is the basis for trading the level of privacy and the quality of service. Since local POIs are more relevant, communication can be limited and hence (trajectory) privacy remains protected (see also Sect. 4.3.2.3).

4.3.3.3 Latency

Latency refers to the length of the delay between when an event occurs and when that event is correctly detected by an algorithm (**P19**. Both et al. 2013; Duckham 2012). As the title suggests, in Laube et al. (**P12**. 2011) on "Deferred decentralized movement mining" roaming nodes require a latency phase for building up their neighborhood memory. The mobile nodes compensate for their limited spatial perception range through extending their temporal perception range, that is, allow for a latency period before the actual data mining step kicks in. Here, computation is deferred, but it is also possible to defer communication. In Laube et al. (**P6**. 2008; **P12**. 2011), and in Both et al. (**P19**. 2013) roaming nodes are given time to rearrange and recombine communication opportunities. Latency is an implication of *mobility diffusion* since time is required for the information to move through the network.

4.3.3.4 Separation

Another feature of decentralized systems is the separation of tasks which in conventional spatial computing are typically hosted in a central and omniscient computing platform. Apart from the obvious spatial separation of partial spatial computing tasks that are, for example, required for computing a boundary, separation can also be of a functional nature. For instance, in Both et al. (**P19**. 2013) the generation of information is separated from the querying of that information. Quite often information is also separated from the rather limited and constrained nodes. In both the fish in Both et al. (**P19**. 2013) and the flocking agents in Laube et al. (**P6**. 2008) information tokens are separated from their carriers such that information can move beyond the constraints of the individual nodes. Hence, separation clearly is a precondition for *mobility compensation* and *mobility diffusion*.

4.3.3.5 Heuristics

The fundamental question as to whether decentralized spatial computing in general can perform as good as conventional spatial computing, with respect to efficiency, effectiveness, accuracy, or scalability, is still subject of ongoing research (**P7**. Laube et al. 2009). Irrespective of this question, approximation approaches or heuristics can be an appropriate way to address tricky tasks in arguably constrained decentralized environments. For instance, Laube et al. (**P6**. 2008) argue that heuristics are a suitable way of compensating for limited perception of individual sensor nodes. The heuristic extrapolates the presence of a flock, reaching beyond any single node's limited perception range. Duckham (2012) follows up on this initial approach and argues that the apparently simple *nkr*-flock is computationally intractable such that even centralized spatial information systems require heuristics in order to compute such patterns (Duckham 2012).

4.4 Related Work

This section summarizes further related work relevant to the topics covered in this chapter. The chapter then concludes with insights and lessons learned from both the research covered in this chapter and in related work.

Wireless sensor networks and geosensor networks. The textbook "Wireless Sensor Networks—An Information Processing Approach" by Zhao and Guibas (2004) offers an excellent entry point into the wider research area. Most research in wireless sensor networks is concerned with lower-level engineering tasks establishing the system infrastructure and maintaining communication (physical, data link, network, transport layers, Zhao and Guibas 2004). The top-most application level, the focus of this book, still receives less attention. More recently Nittel (2009) has summarized the field's progress in monitoring geographic phenomena and advances in dynamic environmental monitoring. She makes very clear that the key task of geosensor networks is to sense, monitor and track *dynamic* phenomena in real-time in the environment. Clearly, this not only refers to change, but also mobility.

Decentralized spatial computing. Duckham's recent textbook on "Decentralized Spatial Computing—Foundations of Geosensor networks" offers an accessible and comprehensive introductory overview on various aspects of DeSC, including many aspects not covered in this book (Duckham 2012). Beyond movement analysis, DeSC presents a new way of processing spatio-temporal information, contrasting to conventional spatial computing in omniscient centralized spatial information systems and databases.

Movement in decentralized spatial information systems. There are more and more integrated spatial systems consisting of mobile networks of computing and sensing platforms: Examples include robot swarms (McLurkin 2008), vehicular ad-hoc networks VANETs (Kosch et al. 2006), mobile phone networks (Ahas et al. 2010), emergency dispatch (Kim et al. 2008), vehicle navigation, or fleet management applications (Arampatzis et al. 2005). However, in most systems the conventional centralized approach for processing and analyzing data prevails.

Decentralized movement analysis principles. There are, however, exceptions aiming at explicitly decentralized ways of analyzing movement that are relevant in the context of this chapter. Shared-ride trip planning systems, for example, are proposed to function on a peer-to-peer basis, requiring a certain degree of in-network data processing for allocating passengers to rides (Dillenburg et al. 2002; Winter and Nittel 2006; Wu et al. 2007). Other examples can be found in the field of robotics, where minimalist robotic swarms are tasked for inspection, maintenance, and repair (Correll and Martinoli 2006; McLurkin 2008). Here, the system relies on roaming mobile nodes addressing a task in a collaborative manner. The work by Grossglauser and colleagues on mobility diffusion played a pivotal role for the development of algorithms for decentralized flock detection (Grossglauser and Tse 2002; Grossglauser and Vetterli 2006). Aiming at improved communication in networks of roaming sensor nodes, these authors propose communication and routing protocols involving (i) the maintenance of local databases that (ii) successively refine their knowledge while moving and through using diffused information. Both concepts are found in a similar way in Laube et al. (**P12**. 2011) adapted for the task of decentralized movement pattern mining.

Finally, the reader is directed to further neighboring areas that may offer relevant concepts and principles for decentralized movement analysis. First, there is a large body of literature on the related but different topic of *distributed data mining in peer-to-peer networks* (Datta et al. 2006; Kargupta and Chan 2000). Note that distributed computing is less constrained than decentralized computing, such that the cooperating systems also synchronously address a computing task but there may be a controlling system part that has access to the entire system state. Second, the textbook by Giannotti and Pedreschi (2008) offers an access point for privacy issues. Apart from technical questions around privacy, the following articles also investigate ethical issues and reflect on lessons learned after a decade of LBS (Dobson and Fisher 2003; Uteck 2009; Nouwt 2008).

4.5 Concluding Remarks

ICT increasingly pervades our dynamic natural and built environments. Decentralized movement analysis results from the application of decentralized spatial computing (DeSC) concepts for CMA. I argue in this chapter that decentralization offers one strategy for coping with emerging big data streams. Initial work reflected in this book and observed in related work indicates the integration of DeSC and CMA and puts forward promising concepts for coping with big data emerging ubiquitous spatial information systems, that in most cases involve some form of mobility.

Movement poses a set of additional challenges to DeSC, but also offers unique opportunities for handling big data streams emerging from dynamic ubiquitous spatial systems. Movement means constantly changing network and neighborhood structures and temporarily broken communication links. However, mobile nodes can carry around information tokens and overcome unfavorable node constellations. Most typical reasons given for DeSC in the first place (Duckham 2012), also hold for decentralized movement analysis scenarios. Local computing reduces information overload in flooded sinks and safeguards user privacy. Mobile systems can be very flexible, resilient and scalable, since added nodes easily extend systems that must grow. Latency allows nodes to explore spatial variables and exchange and enrich the captured information.

Just as in DeSC scenarios in general, decentralized movement analysis problems challenge system and algorithms designers with very peculiar limitations and constraints. The work reported on in this volume in many cases faced such challenges by trading benefits in one aspect for costs in another. Algorithms traded a limited spatial perception versus temporal perception, quality of service versus level of privacy, error of omission versus error of commission, latency versus detection error, or computational complexity versus latency. In some application contexts it is perfectly acceptable to get a task done only after a given latency period or it is equally acceptable to get *a* result, possibly even only an approximation, instead of the best possible, exact result. However, the question whether or not decentralized movement analysis as a form of DeSC can perform just as well as any centralized system remains on open research question.

References

Ahas, R., Silm, S., Järv, O., Saluveer, E., & Tiru, M. (2010). Using mobile positioning data to model locations meaningful to users of mobile phones. *Journal of Urban Technology, 17*(1), 3–27.

Arampatzis, T., Lygeros, J., & Manesis, S. (2005). A survey of applications of wireless sensors and wireless sensor networks. In *Proceedings of the 2005 IEEE International Symposium on Intelligent Control and Mediterrean Conference on Control and Automation*, pp. 719–724.

Augusto, J. C., & Shapiro, D. (Eds.). (2007). *Advances in Ambient Intelligence, Volume 164 of Frontiers in Artificial Intelligence and Applications*. Amsterdam, NL: IOS Press.

Both, A., Duckham, M., Laube, P., Wark, T., & Yeoman, J. (2013). Decentralized monitoring of moving objects in a transportation network augmented with checkpoints. *The Computer Journal*, *56*(12), 1432–1449. doi:10.1093/comjnl/bxs117.

Correll, N., & Martinoli, A. (2006). Collective inspection of regular structures using a swarm of miniature robots. In J. Ang, J. Khatib, & O. Khatib (Eds.), *Experimental Robotics IX, The 9th International Symposium on Experimental Robotics (ISER)*, Singapore, June 18–21 (Vol. 21, pp. 375–385). Springer Tracts in Advanced Robotics. Berlin: Springer.

Datta, S., Bhaduri, K., Giannella, C., Kargupta, H., & Wolff, R. (2006). Distributed data mining in peer-to-peer networks. *IEEE Internet Computing*, *10*(4), 18–26.

Dillenburg, J. F., Wolfson, O., & Nelson, P. C. (2002). The intelligent travel assistant. In *The IEEE 5th International Conference on Intelligent Transportation Systems*, pp. 691–696.

Dobson, J. E., & Fisher, P. F. (2003). Geoslavery. *IEEE Technology and Society Magazine*, *22*(1), 47–52.

Dredge, S. (2013). Waze and means: Google tipped to beat apple and facebook to $1.3bn acquisition. The Guardian.

Duckham, M. (2012). *Decentralized Spatial Computing, Foundations of Geosensor Networks*. Berlin: Springer.

Duckham, M., & Bennett, R. (2009). Ambient spatial intelligence. In B. Gottfried & H. Aghajan (Eds.), *Behaviour Monitoring and Interpretation—BMI—Smart Environments*. Ambient Intelligence and Smart Environments (Vol. 3, pp. 319–335). Amsterdam, NL: IOS Press.

Duckham, M., Nittel, S., & Worboys, M. (2005). Monitoring dynamic spatial fields using responsive geosensornetworks. In C. Shahabi & O. Boucelma (Eds.), *ACM GIS* (pp. 51–60). New York: ACM Press.

Galton, A. (2004). Fields and objects in space, time, and space-time. *Spatial Cognition and Computation*, *4*(1), 39–68.

Giannotti, F., & Pedreschi, D. (2008). Mobility, data mining and privacy: A vision of convergence. In F. Giannotti & D. Pedreschi (Eds.), *Mobility, Data Mining and Privacy* (pp. 1–11). Berlin: Springer.

Greenfield, A. (2006). *Everyware: The dawning age of ubiquitous computing*. Berkeley: New Riders Press.

Grenon, P., & Smith, B. (2004). SNAP and SPAN: Towards dynamic spatial ontology. *Spatial Cognition and Computation*, *4*(1), 69–103.

Grossglauser, M., & Tse, D. N. C. (2002). Mobility increases the capacity of ad hoc wireless networks. *IEEE/ACM Transactions on Networking*, *10*(4), 477–486.

Grossglauser, M., & Vetterli, M. (2006). Locating mobile nodes with ease: Learning efficient routes from encounter histories alone. *IEEE/ACM Transactions on Networking*, *14*(3), 457–469.

Kargupta, H., & Chan, P. (2000). *Advances in distributed and parallel knowledge discovery*. Menlo Park, CA: AAAI Press, the MIT Press.

Kellerer, W., Bettstetter, C., Schwingenschlogl, C., Sties, P., & Steinberg, K. E. (2001). (Auto) mobile communication in a heterogeneous and converged world. *IEEE Personal Communications*, *8*(6), 41–47.

Kim, S., Maciejewski, R., Ostmo, K., Delp, E. J., Collins, T. F., & Ebert, D. S. (2008). Mobile analytics for emergency response and training. *Information Visualization*, *7*(1), 77–88.

Koehn, J., Nicol, S., McKenzie, J., Lieschke, J., Lyon, J., & Pomorin, K. (2008). Spatial ecology of an endangered native australian percichthyid fish, the trout cod maccullochella macquariensis. *Endangered Species Research*, *4*(1–2), 219–225.

Kosch, T. Adler, C. J., Eichler, S. Schroth, C. & Strassberger, M. (2006). The scalability problem of vehicular ad hoc networks and how to solve it. *Wireless Communications, IEEE*, *13*(5), 22–28.

Laube, P., & Duckham, M. (2009). Decentralized spatial data mining for geosensor networks. In H. Miller & J. Han (Eds.), *Geographic data mining and knowledge discovery* (2nd ed., pp. 409–430). London: CRC Press.

Laube, P., Duckham, M., & Croitoru, A. (2009). Distributed and mobile spatial computing. *Computers, Environment and Urban Systems*, *33*(2), 77–78.

Laube, P., Duckham, M., & Palaniswami, M. (2011). Deferred decentralized movement pattern mining for geosensor networks. *International Journal of Geographical Information Science*, 25(2), 273–292. doi:10.1080/13658810903296630.

Laube, P., Duckham, M., & Wolle, T. (2008). Decentralized movement pattern detection amongst mobile geosensor nodes. In T. J. Cova, K. Beard, M. F. Goodchild, & A. U. Frank (Eds.), *Geographic Information Science* (pp. 199–216). Volume 5266 of Lecture Notes in Computer Science. Berlin: Springer. ISBN 978-3-540-87472-0.

Laube, P., Duckham, M., Worboys, M., & Joyce, T. (2010). Decentralized spatial computing in urban environments. In B. Jiang & X. Yao (Eds.), *Geospatial analysis and modelling of urban structure and dynamics* (pp. 53–74). Berlin Heidelberg: GeoJournal Library, Springer.

Lynch, N. (1996). *Distributed algorithms*. San Mateo, CA: Morgan Kaufmann.

McLurkin, J. (2008). Analysis and implementation of distributed algorithms for multi-robot systems (PhD thesis, Massachusetts Institute of Technology).

Nittel, S. (2009). A survey of geosensor networks: Advances in dynamic environmental monitoring. *Sensors*, 9(7), 5664–5678.

Nittel, S., Stefanidis, A., Cruz, I., Egenhofer, M. J., Goldin, D., Howard, A., et al. (2004). Report from the first workshop on geo sensor networks. *ACM SIGMOD Record*, 33(1), 141–144.

Nouwt, S. (2008). Reasonable expectations of geo-privacy? *SCRIPTed*, 5(2), 375–403.

Rule, J., McAdam, D., Stearn, L., & Uglow, D. (1980). *Politics of privacy*. New York: New American Library.

Smith, P., Hutchison, D., Sterbenz, J. P. G., Schöller, M., Fessi, A., Karaliopoulos, M., et al. (2011). Network resilience: A systematic approach. *IEEE Communications Magazine*, 49(7), 88–97.

Uteck, A. (2009). Ubiquitous computing and spatial privacy, anonymity, privacy and identity in a networked society. In I. Kerr, V. Steeves & C. Lucock, (Eds.), *Lessons from the identity trail* (pp. 83–102). Oxford: Oxford University Press.

Werner-Allen, G., Lorinez, K., Ruiz, M., Marcillo, O., Johnson, J., Lees, J., et al. (2006). Deploying a wireless sensor network on an active volcano. *IEEE Internet Computing*, 10(2), 18–25.

Wilensky, U. (1999). Netlogo (and netlogo user manual).

Winter, S., & Nittel, S. (2006). Ad hoc shared-ride trip planning by mobile geosensor networks. *International Journal of Geographical Information Science*, 20(8), 899–916.

Worboys, M., & Duckham, M. (2004). *GIS: A computing perspective* (2nd ed.). New York: CRC Press.

Wu, Y. H., Guan, L. J., & Winter, S. (2007). Peer-to-peer shared ride systems. In S. Nittel, A. Labrinidis, & A. Stefanidis (Eds.), *Advances in geosensor networks* (Vol. 4540). Lecture Notes in Computer Science. Berlin: Springer.

Zhao, F., & Guibas, L. J. (2004). *Wireless sensor networks: An information processing approach*. San Francisco, CA: Morgan Kaufmann Publishers.

Chapter 5
Grand Challenges in Computational Movement Analysis

This final chapter addresses the prospect of Computational Movement Analysis (CMA) as a relatively young research field. The first decade of CMA was shaped by significant technological developments resulting in much increased availability of fine-grained movement data, an innocent and somewhat naïve enthusiasm over moving points resulting in a wide but fragmented variety of methods for movement analysis, and finally due to this lack of a unifying theory of CMA only moderate success in overcoming GIS' and GIScience' legacy of static cartography. The final chapter concludes this book by proposing a set of grand challenges of CMA. The grand challenges arise from observations about current trends in CMA and my personal view on where this young and important research field will develop in the years to come.

This book argues that CMA is an emerging research field with ample momentum witnessing rapid development in many related research fields and application areas. Preparing the discussion of the *grand challenges* of the field to follow in the next section, it is here worth back-pedaling for a moment, and trying to capture the current state of a field seemingly moving forwards with giant steps. The following list gives an overview of trends and developments that arguably have and will have implications for the further development of the young research field.

- Movement data becomes more available. Scientists from a wide range of application fields are more willing to share their data.
- Following miniaturization and reduced costs, movement data sets cover more and more individuals at ever finer granularities.
- The movement data sets typically used for CMA change their character: Large repositories capturing trajectory data of massive sets of individuals as a "byproduct" (as for example form GSM networks) outplay purposefully collected movement data sets tracking small numbers of individuals (most movement ecology studies tracking samples sizes of a couple of dozens).
- Data repositories simplifying data exchange amongst scientists are established (for example, http://www.movebank.org) and gain momentum.

© The Author(s) 2014

P. Laube, *Computational Movement Analysis*,
SpringerBriefs in Computer Science, DOI 10.1007/978-3-319-10268-9_5

- Difficulties in bringing theory and application researchers together prevail. Some application areas (for example, movement ecology, crowd management) maintain and further develop CMA research fields with ample momentum but little links to GIScience and computer science.
- Multi-sensor systems (speed, acceleration, physical properties) replace pure location tracking systems.
- After a cooler phase, privacy recovers as a key CMA topic.

After the golden first decade of CMA, progress seems to be slowing down as the low hanging fruit are gone and the hard problems prevail. Even though movement analysis is still prominent on most schedules of relevant conferences, I would argue that the research field gains breadth rather than depth. There seems to be no bound to the arrival of new and and fascinating movement data sources, each with a specific problem producing a unique movement analysis solution. Even though any growth benefits the field, if the field wants to mature, a set of grand challenges need to be addressed.

Clearly, aspects of the following relate to old and well known challenges of handling spatio-temporal data, such as scale, uncertainty, or data integration. Other aspects of CMA pose new and interesting challenges, opening up countless research avenues for the years to come.

5.1 Coping with Big Movement Data

GIScience and geomatics are currently experiencing an exciting revolution, that also has implications for CMA. What started off as a data poor and computation poor discipline, then coped with data rich and computation rich environments for decades, is now faced with another dramatic shift regarding its data sources: Big data (Graham and Shelton 2013; Kitchin 2013) Sensor networks and mobile GIS, user-generated content and open data, the web and cloud computing inevitably change how we capture and manage geoinformation, how we analyze and exploit geoinformation, and ultimately how we take decisions based on geoinformation.

These new and massive data streams not only challenge CMA in terms of volume, but also require new ways of integrating heterogeneous multi-source information (Fig. 5.1). Such integration could, for example, require inferring behavior from GPS tracks sampled at one second intervals, combined with accelerometer readings that come in 6 s long bursts sampled at 20 Hz, but only every 10 min, and interpolated and hence uncertain meteorological field data with a spatio-temporal granularity of daily values per 1 km grid cells. Big geodata as a source for CMA means inferring knowledge and making decisions based on more comprehensive but at the same time more uncertain, messy and noisy movement data. If and to what degree big data really poses fundamentally new challenges is still widely discussed. One could argue that especially GIScience with its long tradition of handling voluminous and messy data is in an excellent position for contributing to the big data debate.

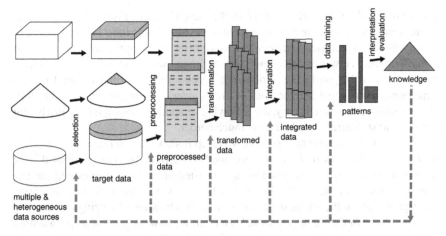

Fig. 5.1 The KDD process and its core step data mining in a world of big geo-data, adapted from (Fayyad et al. 1996)

Mayer-Schönberger and Cukier (2013) emphasize three core challenges around big data. Firstly, the authors argue that the need to sample is an artifact of an era dominated by information scarcity. Overcoming this information scarcity requires a fundamental shift in analytical thinking, where analysis not primarily bases on samples but moves more and more towards analyzing entire (statistical) populations. For example, the group around Rein Ahas has, as of 2013, access to mobile phone logs of almost the entire Estonian population through building up cooperations with all major service providers. Secondly, traditional ways of reasoning and analysis must be complemented with new approaches inferring knowledge from more comprehensive but at the same time much more uncertain and noisy, in short, messy data. Mayer-Schönberger and Cukier (2013) argue that gaining access to more comprehensive data sets allows shedding some of the rigid exactitude required when analyzing conventional samples. Clearly, positional uncertainty will remain an issue with tracking data for many years to come. Compared to the long and intense history of research on uncertainty, it is surprising to see so few articles addressing this pivotal problem in CMA. Thirdly, correlations found in big data may not allow us to understand *why* something is happening (causality), but exploit the correlations found for the equally important alert *that* something is happening.

5.2 Bridging the Semantic Gap

CMA aims at bridging the gap between low-level movement data and the high-level conceptual schemes required for understanding movement processes, just as it must be the goal for GIScience in general (Galton 2005). Through the work summarized in this book and further related work in the area, GIScience has made significant progress in seeking structure in movement data. Algorithms have been proposed to cluster similar trajectories and segment trajectories, to compute home ranges and to find patterns. However, attaching such structures to colloquial names ("flock", "herd", "convoy", "single file"), does not necessarily mean that a found pattern corresponds to an actual herd of cows or trucks indeed moving in a convoy. In short, whereas finding structure is easy, the following semantic annotation and enrichment of the found structures is and remains much harder.

Progress in this second but crucial CMA step is slow for two main reasons. First, most work so far was tools-driven and methods-driven, but not problem-driven. Second, most movement data used so far consisted of bare trajectories without any form of semantic metadata. For instance, when cows were tracked, no information was captured about the social structure of animals. Or, when observing two users of a mobile phone in the same cell for half an hour, we have in general no information knowing for certain that they actually met for coffee. That means, even when aiming at a semantic evaluation of proposed methods, finding appropriate data to do so is difficult. Nevertheless, when aiming at really understanding the processes and events controlling the observed movement, then bridging the semantic gap between formal representations of patterns and structures and their actual meaning grounded in contextual expert knowledge is key. Initial work combining movement data with social media data (for example, applications allowing users to "check-in" at points of interest) offer a promising route for enriching raw trajectories with user-generated semantics (Sui and Goodchild 2011).

Some research fields concerned with CMA have proposed promising work addressing this gap. Regarding data capture, recent work in movement ecology no longer just monitors the location of observed animals, but uses multi-sensor devices simultaneously tracking physical properties of the individuals (for example, speed, acceleration, physiological variables). The underlying rational is that activities can't be inferred from location fixes alone, irrespective of how densely they are sampled. A second development in CMA specifically aims at understanding movement by understanding its embedding in its enabling and constraining geospatial context (**P3**. Laube et al. 2007). Cows that are co-located for some time may not belong to the same herd at all but just be kept in a fenced area. Similarly, understanding commuter patterns without studying the transportation infrastructure obviously makes little sense. Again, given the strong influences of geometry and topology oriented researchers, it may be little surprising that so far most CMA focuses on shape and arrangement of trajectories and less on the development of context-aware movement analysis techniques.

5.3 Contributing to Ambient Spatial Intelligence

Ever smarter smart phones and prospects like Google Glass change the public perception of spatial information. Using spatial information became as common as writing
an SMS or downloading one's favorite music—anywhere, any time. This (brave)
new world is a very dynamic world. Not only do highly mobile human users interact in dynamic networks and profit from ubiquitous access to spatial information,
but spatially distributed autonomous computing nodes increasingly invade various
application fields of immense socio-economic significance, including applications
in ICT, transportation, and logistics.

In this world of "spatial everyware", the days of spatial data processing in a
monolithic desktop GIS are numbered. Today's and forward-looking CMA algorithms must comply with highly dynamic multiparty networks, where data volumes,
privacy issues, and constantly changing communication networks dictate that spatial
data collected at different sites be analyzed in a decentralized way without collecting
all data to a central GIS or database. In such distributed and integrated systems the
boundaries between data capture and data processing are increasingly blurred, nurturing the vision of ambient spatial intelligence. CMA has an opportunity here by
contributing in an early stage to this emerging field through a combination of classic
GIScience strengths with forward-looking decentralized spatial computing.

5.4 Balancing Benefits and Privacy

Safeguarding user privacy remains a technical challenge and a key ethical responsibility for researchers in CMA. Clearly, animal movement is an excellent use case
for stimulating and interesting CMA problems. But the movement of people and the
related applications and services bears a much larger socio-economic potential. It is
here where CMA must seek its primary contribution, this is a huge opportunity not
to be missed. But people care about privacy. I see two main challenges with respect
to privacy: First, develop strategies for getting access to the really interesting large
volumes of people movement data. Second, develop analytical frameworks that can
produce useful information but at the same time safeguard peoples' privacy.

Access to GSM network data is still very limited, studies where access to very
large numbers of individuals is granted are still rather the exception or don't hap-
pen in the public or scientific domain. Clearly, information and transparency about
a study's goals and privacy precautions help building up trust with users and GSM
providers. Rein Ahas' Estonian case may serve as a successful example here. An
alternative promising strategy for accessing trajectories of large numbers of individuals comes in the form of apps that track their users in informed consent in some
application context. For example, Wirz et al. (2012) propose a system for real-time
crowd monitoring where a mobile phone app is used that supplies the user with
event-related information, but periodically logs the device's location along the way.

Here, the users of the app reveal their location by giving informed consent, in turn receiving information about the event they are visiting.

Whereas some promising solutions have been proposed for safeguarding user privacy in location-based services (LBS), in applications where the analysis of movement data is the primary goal, much less convincing privacy handling prevails. CMA must devise analytical frameworks that allow the inference of useful knowledge but at the same time safeguard the privacy of the tracked people. Strategies include *anonymity*, *spatial* or *temporal degradation*, or *delay* (Krumm 2009). Similar to Laube et al. (**P11**. 2010) where privacy was traded for quality of service in a LBS scenario, CMA must develop techniques where quality of insights can be balanced with the level of privacy.

5.5 Improving Recognition

So far, CMA remains a bit a "scientists' science", comparable to modal jazz sometimes being called "musicians' music", that is, music mainly for musicians and hence inaccessible to non-musicians. Even though several community activities, including a series of workshops[1] and even a EU-funded COST action (COST Action IC0903 MOVE) specifically aimed at bringing together methods and application scientists, the prevailing pattern is that application experts from government and industry rarely participate. Clearly, this is more than a challenge, this is an urgent problem to be solved.

One way of improving the visibility and recognition of CMA as a research field, is seeking publication of CMA work not only in GIScience outlets but also in related fields such as core computer science, ecology, or transportation research. There is no point in complaining that the established GIScience theory is ignored by computer scientists "reinventing GIS" in the course of the rapid development of mobile ICT and location-based apps. It is the responsibility of researchers active in GIScience (and hence CMA) to seek the widest possible visibility. Clearly, close collaboration with problem-driven application experts helps in producing work appealing to a wider audience.

Second, and perhaps more difficult, is the establishment of one or two killer applications underlining the socio-economic relevance of CMA. This is surely more difficult as such success also depends on external constraints (for example, the established GIScience concept LBS only became a commercial success when app-stores became popular). Nevertheless, again only targeted collaboration on application-close problems have the potential to produce such relevance in the first place.

---

[1] Dagstuhl Seminars #08451 (2008), #10491 (2010) #12512 (2012), on Representation, Analysis and Visualization of Moving Objects; First Workshop on Movement Pattern Analysis (MPA'10), 09/2010, Zurich, Switzerland; Workshop on Analysis and Visualization of Moving Objects, Lorentz Centre, 06/2011, Leiden, NL; Workshop on Progress in Movement Analysis—Experiences with Real Data, 09/2012, University of Zurich, Switzerland.

5.6 Towards a Unifying Theory of CMA

Progress in the above challenges is much more likely when CMA manages to establish a unifying theory. This book contributes to the establishment of such a unifying theory by providing a structured overview of concepts, techniques and their implications for three important CMA aspects. As an umbrella research field, CMA must persistently continue its efforts to agree on ontological foundations and common basic tasks, and intensify the sharing of data and methods. Initial community efforts in that direction sparked the flame but must now be followed up. When researchers from the many contributing fields manage to bundle their efforts, CMA can make a real contribution to the present and future challenges of a world in motion.

References

Fayyad, U., Piatetsky-Shapiro, G., & Smyth, P. (1996). From data mining to knowledge discovery in databases. *AI Magazine, 17*(3), 37–54.

Galton, A. (2005). Dynamic collectives and their collective dynamics. In A. Cohn & D. M. Mark (Eds.), *Spatial information theory, proceedings*, (Vol. 3693, pp. 300–315)., Lecture Notes in Computer Science Berlin: Springer.

Graham, M., & Shelton, T. (2013). Geography and the future of big data, big data and the future of geography. *Dialogues in Human Geography, 3*(3), 255–261.

Kitchin, R. (2013). Big data and human geography: Opportunities, challenges and risks. *Dialogues in Human Geography, 3*(3), 262–267.

Krumm, J. (2009). A survey of computational location privacy. *Personal and Ubiquitous Computing, 13*(6), 391–399.

Laube, P., Dennis, T., Walker, M., & Forer, P. (2007). Movement beyond the snapshot—dynamic analysis of geospatial lifelines. *Computers, Environment and Urban Systems, 31*(5), 481–501.

Laube, P., Duckham, M., Worboys, M., & Joyce, T. (2010). Decentralized spatial computing in urban environments. In B. Jiang & X. Yao (Eds.), *Geospatial analysis and modelling of urban structure and dynamics, geojournal library* (pp. 53–74). Berlin: Springer.

Mayer-Schönberger, V., & Cukier, K. (2013). *Big data: A revolution that will transform how we live, work, and think*. Boston, MA: Houghton Mifflin Harcourt.

Sui, D., & Goodchild, M. (2011). The convergence of gis and social media: challenges for giscience. *International Journal of Geographical Information Science, 25*(11), 1737–1748.

Wirz, M., Franke, T., Roggen, D., Mitleton-Kelly, E., Lukowicz, P., & Troster, G. (2012). Inferring crowd conditions from pedestrians' location traces for real-time crowd monitoring during city-scale mass gatherings. *IEEE 21st International Workshop on Enabling Technologies: Infrastructure for Collaborative Enterprises (WETICE)*, pp. 367–372.